CONCEPTS AND TECHNIQUES IN MODERN GEOGRAPHY No. 8

PRINCIPAL COMPONENTS ANALYSIS

by

Stu Daultrey

(University College Dublin)

CONTENTS

		Page
I	THE TECHNIQUE DESCRIBED	
(i)	Introduction	3
(ii)	Prior knowledge assumed	4
(iii)	Conventions and notations	4
(iv)	Correlation	5
II	THE TECHNIQUE EXPLAINED	
(i)	The transformation of one co-ordinate system to another	10
(ii)	The derivation of the principal components	15
(iii)	Principal component loadings	17
(iv)	Principal component scores	18
(v)	Further reading	18
(vi)	A worked example	19
(vii)	The significance of principal components	24
III	THE TECHNIQUE APPLIED	
(i)	Introduction	26
(ii)	The data	27
(iii)	The first analysis	30
(iv)	The second analysis	35
IV	THE TECHNIQUE ELABORATED	
(i)	The normal distribution and the principal components of correlation matrices	41
(ii)	Principal components analysis of dispersion matrices	42
(iii)	Discarding variables	42
(iv)	Rotation of principal components	44

(v)	The analysis of component scores	44
(vi)	Sampling properties of principal components	46
(vii)	Conclusions	47
	BIBLIOGRAPHY	48

Acknowledgement

The author would like to thank Jim Walsh of McMaster University for permission to use the data presented and analysed in chapter 3; and Brendan Quigley (UCD) and Willie Smyth (Maynooth) for their constructive comments on matters mathematical and agricultural respectively. Responsibility for the final text is, of course, the author's.

I THE TECHNIQUE DESCRIBED

(i) Introduction

Principal components analysis is a data transformation technique. If, for a series of sites, or objects, or persons, a number of variables is measured, then each variable will have a variance (a measure of the dispersion of values around the mean), and usually the variables will be associated with each other, i.e. there will be covariance between pairs of variables. The data set as a whole will have a total variance which is the sum of the individual variances.

Each variable measured is an axis, or dimension, of variability. What happens in principal components analysis is that the data are transformed to describe the same amount of variability, the total variance, with the same number of axes, the number of variables, but in such a way that:

- the first axis accounts for as much of the total variance as possible;
- the second axis accounts for as much of the remaining variance as possible whilst being uncorrelated with the first axis;
- the third axis accounts for as much of the total variance remaining after accounted for by the first two axes, whilst being uncorrelated with either;
- and so on.

The new axes, or dimensions, are uncorrelated with each other, and are weighted according to the amount of the total variance that they describe. Normally this results in there being a few large axes accounting for most of the total variance, and a larger number of small axes accounting for very small amounts of the total variance. These small axes are normally discounted from further consideration, so that the analyst has transformed his data set having p correlated variables to a data set that has m uncorrelated axes, or principal components, where m is usually considerably less than p.

The fact that the m axes are uncorrelated is often a very useful property if further analysis is planned. The fact that m is less than p introduces a parsimony which is more often than not desirable in any scientific discussion. But much attention focusses on the relationship of the principal components to the original variables - which of the axes of variability contributed most variance to each of the principal components? Put another way, how can each principal component be interpreted in terms of the original variables? And finally, given that each site, object, or person had a value of each of the original variables, what are the values, or scores, associated with any one case measured in units of the new axes, the principal components? In other words, what are the values of the data points after the transformation?

After statements about the level of comprehension required of the reader, and an explanation of the notations used throughout the text, the rest of Part I is concerned with correlation between variables, and introduces some hypothetical data which will be used to illustrate the derivation of the principal components.

Part II is concerned with the derivation of the magnitudes and directions of the principal components of a data set, which, once obtained, allow the description of the relation between the original variables and the principal components, and thence the definition of the data points in terms of the principal components.

Part III demonstrates the application of principal components analysis to an actual data set, in this case information on post-War changes in agriculture in the Republic of Ireland.

Some of the problems arising from the analysis in Part III are dealt with in Part IV, together with a more general discussion of principal components analysis in geographical studies.

(ii) Prior knowledge assumed

It is assumed that the reader:

1. can add, subtract, multiply and divide;
2. has taken a course in elementary statistical methods, and is familiar with the descriptive statistics mean, variance, covariance, with probability distributions e.g. the normal distribution, and with hypothesis testing.
3. is acquainted with the rudiments of matrix algebra.

If assumption 3. does not hold then the reader is referred to the excellent synopses and bibliographies in Krumbein & Graybill (1965; chapter 11) and King (1969; appendix A4), Morrison (1967; chapter 2).

(iii) Conventions and notations

As far as possible standard notations have been adhered to. The Greek symbols represent population statistics, whilst ordinary letters represent sample statistics. The exception is the use of λ and Λ to denote eigenvalues whether or not they are derived from sample data; the Greek symbol is used because a typewritten lower case l is easily confused with the number 1 or the letter i. L is used instead to denote the matrix of component loadings.

C is used to denote the sample variance-covariance matrix in preference to the conventional S because the elements of {S} would become $\{s_{ij}\}$ (see below), whereas when referred to individually, the diagonal elements, the variances, are denoted s_i^2; the use of {C} and $\{c_{ij}\}$ avoids all confusion.

The matrix notation is conventional:

Any matrix is {A}, with elements a_{ij} where i refers to the row and j the column. For a 3x3 matrix

$$\{A\} = \begin{Bmatrix} a_{11} & a_{12} & a_{13} \\ a_{21} & a_{22} & a_{23} \\ a_{31} & a_{32} & a_{33} \end{Bmatrix}$$

Any vector is {a}, with elements a_j for a row vector, a_i for a column vector.

The transpose of any matrix is denoted $\{A\}^T$.
The inverse of any matrix is denoted $\{A\}^{-1}$.
The product of two matrices is shown as $\{A\}.\{B\}$.
The determinant of any matrix is written $|A|$, or

$$|A| = \begin{vmatrix} a_{11} & a_{12} & a_{13} \\ a_{21} & a_{22} & a_{23} \\ a_{31} & a_{32} & a_{33} \end{vmatrix}$$

(iv) <u>Correlation</u>

Principal components analysis depends upon the fact that at least some of the variables in the data set are intercorrelated. If none of the p variables is correlated with any other, there exists already a set of uncorrelated axes, and there is no point in performing a principal components analysis.

Correlation is the association between two variables - the amount by which they covary. The most frequently used measure is Pearson's product moment correlation coefficient, which is a parametric measure of linear association. It is defined as the ratio of the covariance between two variables to the square root of the product of the two variances.

$$\rho = \frac{Cov(X_1,X_2)}{(\sigma_1^2 \cdot \sigma_2^2)^{\frac{1}{2}}} = \frac{Cov(X_1,X_2)}{\sigma_1 \cdot \sigma_2} \qquad (1)$$

Working with sample data, r, an unbiassed estimate of the population correlation coefficient ρ, becomes

$$r = \frac{Cov(X_1,X_2)}{s_1 \cdot s_2} = \frac{N\Sigma X_1 X_2 - \Sigma X_1 \cdot \Sigma X_2}{((N\Sigma X_1^2 - (\Sigma X_1)^2) \cdot (N\Sigma X_2^2 - (\Sigma X_2)^2))^{\frac{1}{2}}} \qquad (2)$$

It varies from +1.0 (perfect positive correlation) through zero (no correlation) to -1.0 (perfect negative correlation). If there are two <u>uncorrelated</u> variables, each with a mean of zero and unit variance, with <u>a</u> bivariate normal distribution, then the equiprobability contours (lines enclosing an area where there is a given probability of a value occurring) are circular. The axes interest at $90°$, and $\cos 90° = r = 0$. Equiprobability contours for $\rho = 0.5, 0.75, 0.95, 0.99, 0.999$ are shown in fig. 1.

If the two variables are correlated, however, the contours of equiprobability are elliptic, the ellipse being defined by two axes, or vectors, intersecting at the point (μ_1, μ_2), the cosine of the acute angle of intersection being equal to the correlation coefficient. Fig. 2 shows the same equiprobability contours as does fig. 1, but for the case where $\rho = 0.5$.

Ultimately, when there is perfect correlation, the vectors are at an acute angle to each other of $0°$; the ellipse has contracted to a straight

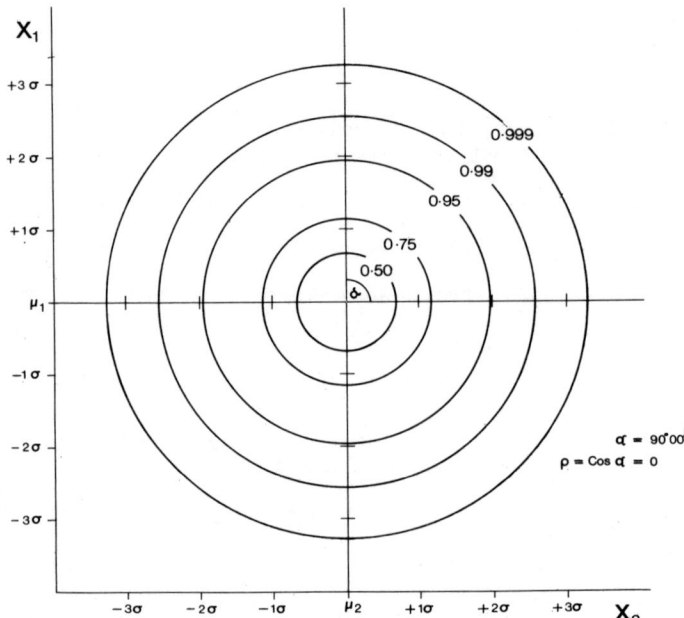

Fig. 1 Equiprobability contours for two uncorrelated variables

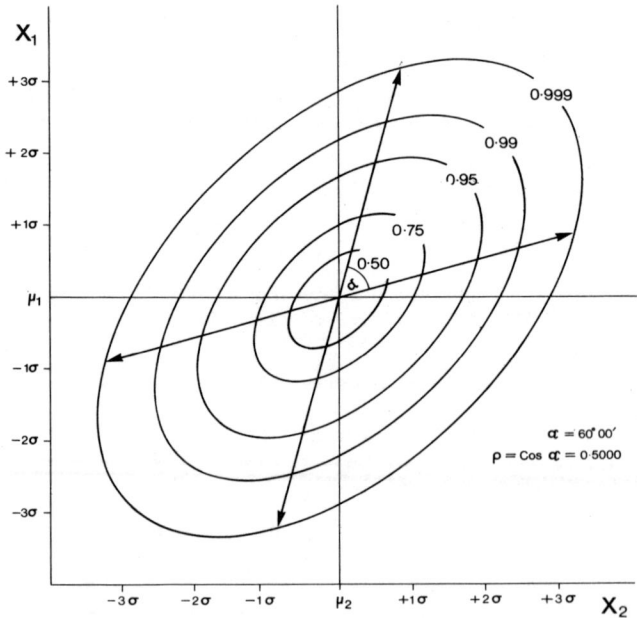

Fig. 2 Equiprobability contours for two correlated variables

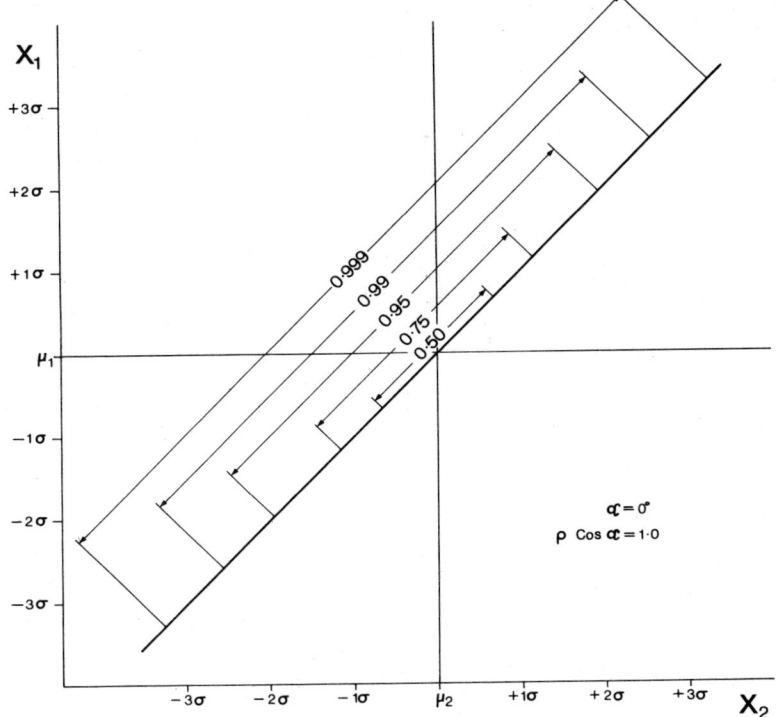

Fig. 3 Equiprobability trace for two perfectly correlated variables

line on which each member of the bivariate normal population lies, and thus one variable is completely defining the other. This case is illustrated in fig. 3.

A simple example of two variable correlation is introduced here, and will be developed later. Nine data points have the following values for two variables, X_1 and X_2.

Data point	X_1	X_2
1.	1	2
2.	2	3
3.	3	1
4.	4	6
5.	5	5
6.	6	4
7.	7	8
8.	8	9
9.	9	7

$\Sigma X_1 = 45$ $\Sigma X_2 = 45$ $\Sigma X_1 X_2 = 275$
$\Sigma X_1^2 = 285$ $\Sigma X_2^2 = 285$ $N = 9$

Table 1. Scores on two hypothetical variables, X_1 and X_2.

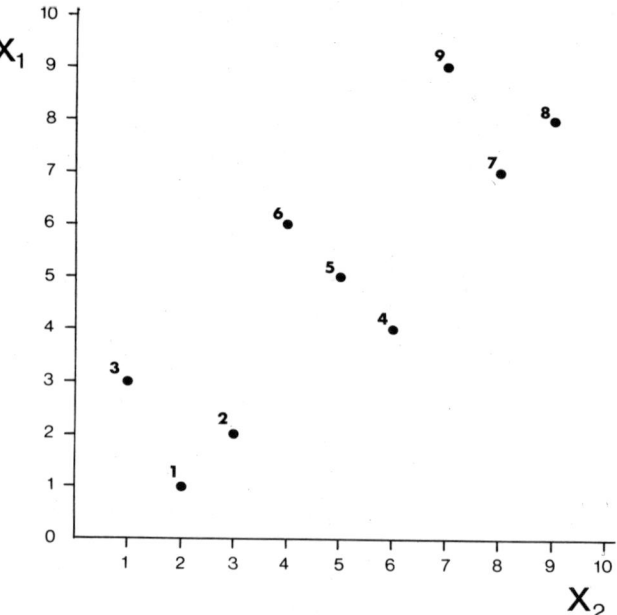

Fig. 4 Plot of hypothetical data for two variables

$$r = \frac{9.275 - 45.45}{((9.285 - 45^2).(9.285 - 45^2))^{\frac{1}{2}}} = \frac{450}{540} = 0.8333$$

Fig. 4 shows the data points; and fig 5 shows for the same data one elliptical equiprobability contour defined by two vectors intersecting at 33° 34' (cos 33° 34' = 0.8333 - r). Fig. 6 shows two orthogonal axes superimposed on the ellipse such that the long axis is as long as possible, and the other axis is as long as possible whilst being at right angles to the first. These axes are the principal components; the long axis is the first component, accounting for as much of the total variance (the area of the ellipse) as possible, and the other axis is the second component, in this case accounting for the remainder of the total variance.

Describing the principal components of two variables is instructive, but substantively trivial; normally, many more than two variables are analysed. It is impossible to draw more than three dimensions, thus in this section and in section II(vi) much of the description will focus on a three variable example.

Just as two uncorrelated variables have circular contours of equiprobability, three uncorrelated variables have spherical equiprobability surfaces (see fig. 7, which shows the p = 0.99 surface). Three <u>correlated</u> variables have ellipsoidal equiprobability surfaces. If a third <u>variable is</u> added to the two in table 1.

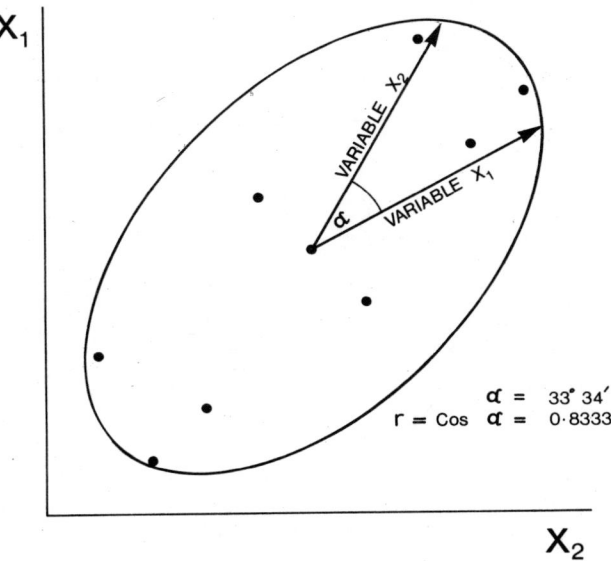

Fig. 5   Hypothetical data with two variables and equiprobability contour

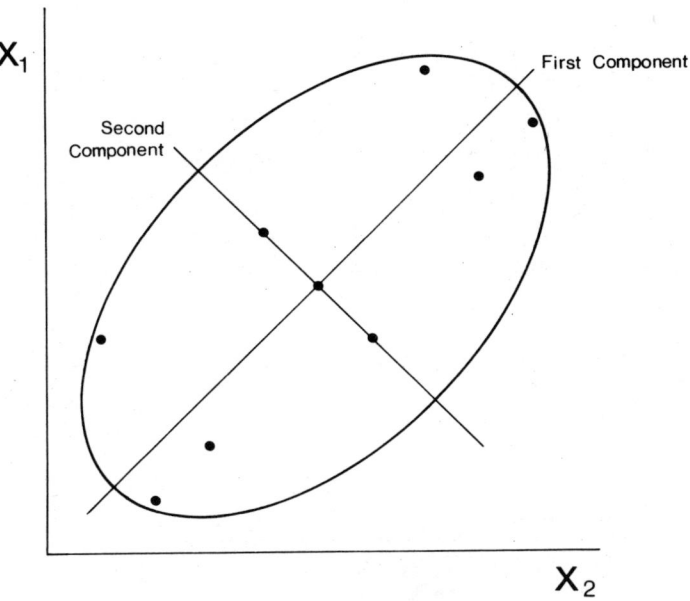

Fig. 6   Hypothetical data with two principal components and equiprobability contour

Data point	X_1	X_2	X_3
1.	1	2	3
2.	2	3	4
3.	3	1	1
4.	4	6	7
5.	5	5	5
6.	6	4	2
7.	7	8	9
8.	8	9	8
9.	9	7	6

$$\Sigma X_1 = \Sigma X_2 = \Sigma X_3 = 45$$

$$\Sigma X_1^2 = \Sigma X_2^2 = \Sigma X_3^2 = 285 \qquad N = 9$$

$$\Sigma X_1 X_2 = 275; \quad \Sigma X_1 X_3 = 260; \quad \Sigma X_2 X_3 = 280$$

Table 2. Scores on three hypothetical variables, X_1, X_2 and X_3.

then a 3 x 3 matrix of correlation coefficients can be calculated.

	X_1	X_2	X_3
X_1	1.0000	0.8333	0.5833
X_2	0.8333	1.0000	0.9167
X_3	0.5833	0.9167	1.0000

Table 3. Matrix of correlations from data in table 1.2.

The data are plotted in fig. 8; one ellipsoidal equiprobability surface, defined by three vectors intersecting at angles whose cosines correspond to the correlation coefficients, is drawn in on fig. 9. Figure 10 shows the three principal components of the ellipsoid, the first being the longest axis through the volume (describing the maximum possible proportion of the total variance), the second the longest possible at right angles to the first (describing the maximum proportion of the remaining variance), and the third the longest possible at right angles to the other two (describing the remaining variance).

II THE TECHNIQUE EXPLAINED

(i) The transformation of one co-ordinate system to another

The problem then, is to define the relationships between the variable axes and the principal components. In a two variable situation the solution is quite simple, and is illustrated in fig. 11. When measured in units of the two variables, a data point D has co-ordinates (X_1, X_2). If the origin, O, is defined as the point (μ_1, μ_2), the co-ordinates of D become $(X_1-\mu_1, X_2-\mu_2)$, or

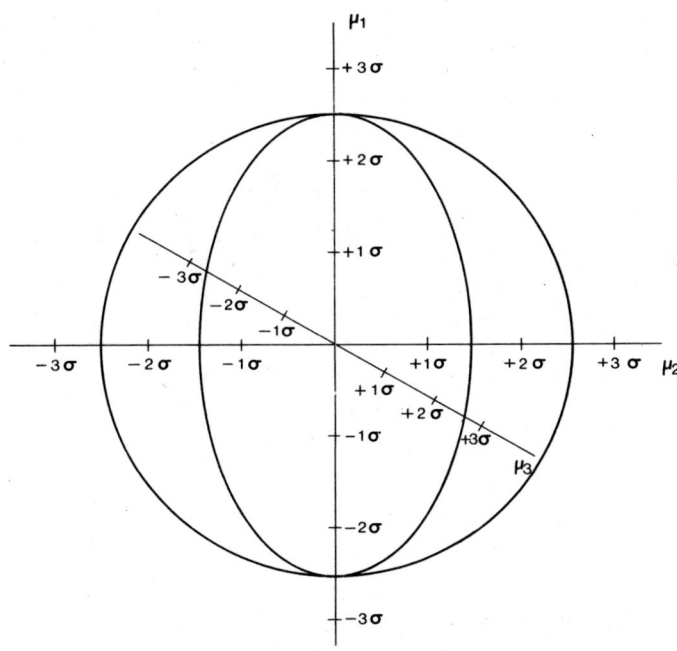

Fig. 7   Spherical equiprobability surface for three uncorrelated variables

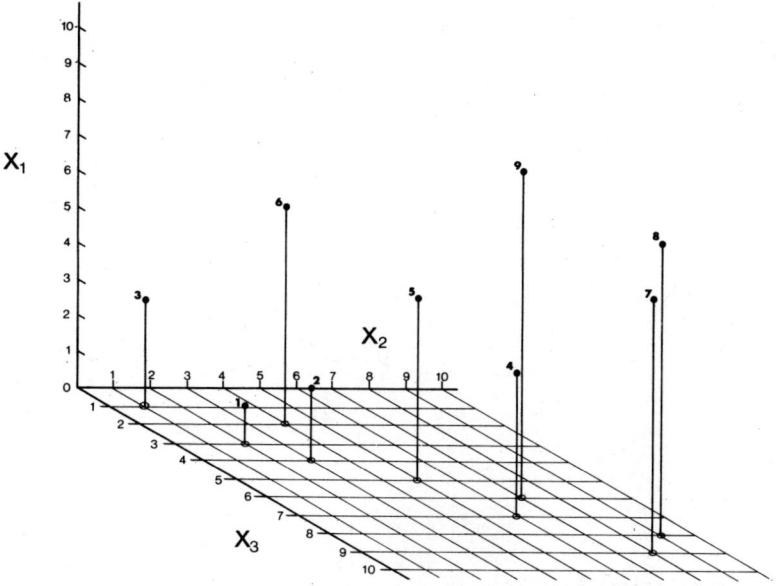

Fig. 8   Plot of hypothetical data for three variables

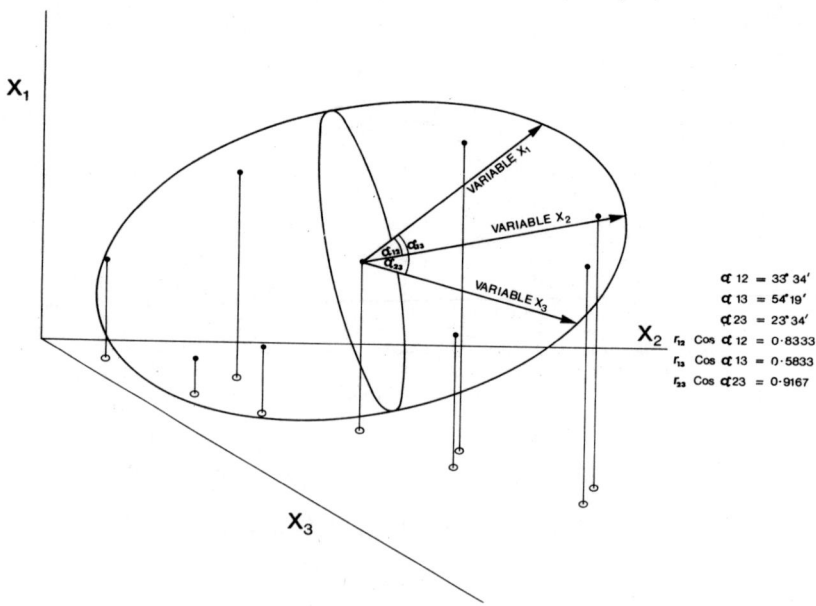

Fig. 9   Hypothetical data with three variables and equiprobability surface

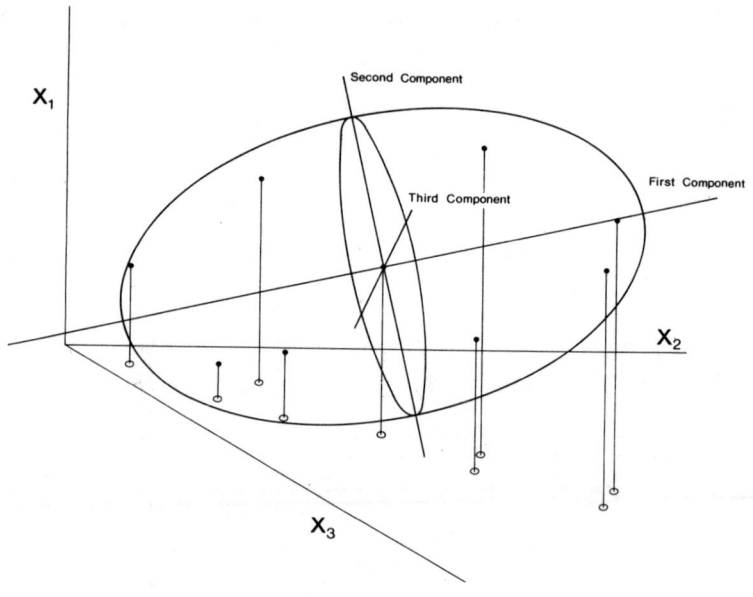

Fig. 10   Hypothetical data with three principal components and equiprobability surface

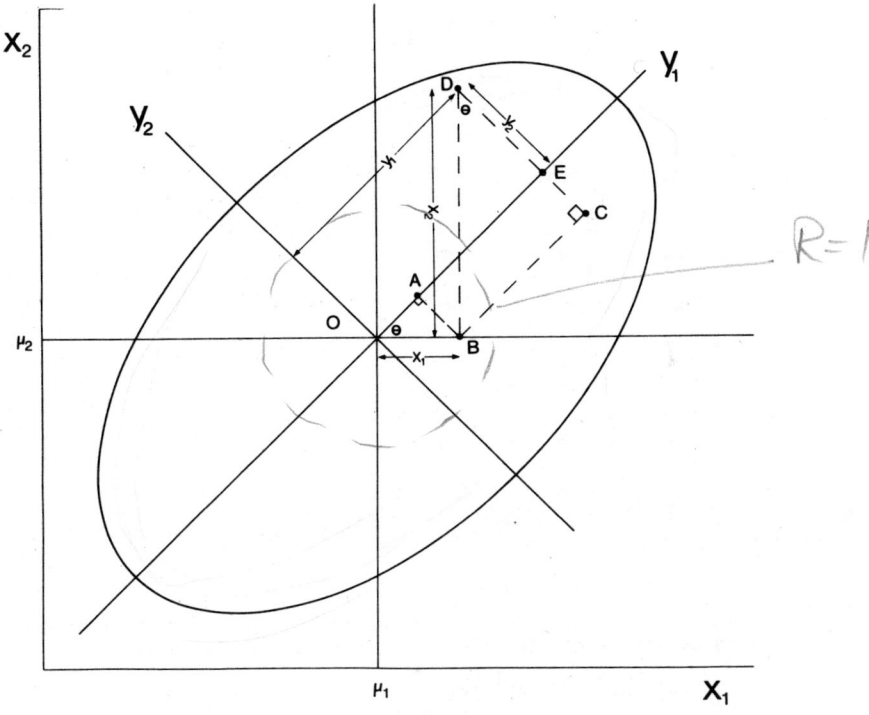

Fig. 11 Transformation of co-ordinate systems for two variables.

in the conventional notation for deviations from the mean (x_1, x_2). When measured in units of the two principal components Y_1 and Y_2, the point D has co-ordinates (y_1, y_2). The relation between the axes X_1, X_2 and Y_1, Y_2 is defined by the angle between them, denoted θ.

$$y_1 = OE = OA + AE \tag{3}$$

$$\cos\theta = \frac{OA}{OB} \quad \therefore \quad OA = OB \cdot \cos\theta = x_1 \cos\theta$$

$$\sin\theta = \frac{BC}{BD} = \frac{AE}{BD} \quad \therefore \quad AE = BD \cdot \sin\theta = x_2 \sin\theta$$

Therefore

$$y_1 = x_1 \cos\theta + x_2 \sin\theta \tag{4}$$

Again
$$y_2 = DE = DC - CE \tag{5}$$

$$\cos\theta = \frac{DC}{BD} \quad \therefore DC = BD\cdot\cos\theta = x_2\cos\theta$$

$$\sin\theta = \frac{AB}{OB} = \frac{CE}{OB} \quad \therefore CE = OB\cdot\sin\theta = x_1\sin\theta$$

Therefore
$$y_2 = x_2\cos\theta - x_1\sin\theta \tag{6}$$

Arranging equations (4) and (6)
$$y_1 = x_1\cos\theta + x_2\sin\theta$$
$$y_2 = -x_1\sin\theta + x_2\cos\theta$$

which is a matrix equation

$$\begin{Bmatrix} y_1 \\ y_2 \end{Bmatrix} = \begin{Bmatrix} \cos\theta & \sin\theta \\ -\sin\theta & \cos\theta \end{Bmatrix} \cdot \begin{Bmatrix} x_1 \\ x_2 \end{Bmatrix} \tag{7}$$

or
$$\{y\} = \{A\}\cdot\{x\} \tag{8}$$

Therefore
$$\{x\} = \{A\}^{-1}\cdot\{y\} \tag{9}$$

As it happens, the matrix $\{A\}$ is orthogonal, which means that when multiplied by its transpose, the product is the identity matrix $\{I\}$.

$$\{A\}\cdot\{A\}^T = \{A\}^T\cdot\{A\} = \{I\} \tag{10}$$

Since the definition of the inverse of a matrix is

$$\{A\}\cdot\{A\}^{-1} = \{A\}^{-1}\cdot\{A\} = \{I\} \tag{11}$$

it follows that the transpose of an orthogonal matrix is equal to its inverse, in this case

$$\{A\}^T = \{A\}^{-1} = \begin{Bmatrix} \cos\theta & -\sin\theta \\ \sin\theta & \cos\theta \end{Bmatrix}$$

Equations (10) and (11) can be verified by

$$\begin{Bmatrix} \cos\theta & \sin\theta \\ -\sin\theta & \cos\theta \end{Bmatrix} \cdot \begin{Bmatrix} \cos\theta & -\sin\theta \\ \sin\theta & \cos\theta \end{Bmatrix}$$

$$= \begin{Bmatrix} \cos^2\theta + \sin^2\theta & -\cos\theta\sin\theta + \sin\theta\cos\theta \\ -\sin\theta\cos\theta + \cos\theta\sin\theta & \sin^2\theta + \cos^2\theta \end{Bmatrix} = \begin{Bmatrix} 1 & 0 \\ 0 & 1 \end{Bmatrix} \tag{12}$$

since $\cos^2\theta + \sin^2\theta = 1$

(ii) **The derivation of the principal components**

If the variance-covariance matrix of the original variables is designated $\{C\}$, then the variance-covariance matrix of the principal components, $\{\Lambda\}$, is

$$\{\Lambda\} = \{A\}.\{C\}.\{A\}^T \tag{13}$$

In the matrix $\{A\}$ the relation between the variables and each principal component is given by the row vectors. However it is more convenient to refer to these relations by the column vectors, so it is necessary to define a matrix $\{E\}$ which is the transpose of $\{A\}$. Equation (13) then becomes

$$\{\Lambda\} = \{E\}^T.\{C\}.\{E\} \tag{14}$$

which accords with the definition of $\{\Lambda\}$ given in King (1969, p.169) and Morrison (1967, p.223), and derived from Anderson (1958, pp 272-279). Because the principal components are uncorrelated, their covariance terms (the off-diagonal elements) are zero, and the matrix $\{\Lambda\}$ is a diagonal matrix; in the two-dimensional case it is

$$\{\Lambda\} = \begin{Bmatrix} \lambda_1 & 0 \\ 0 & \lambda_2 \end{Bmatrix}$$

with the λ_i representing the variances of the two principal components. The principal components account for all the variance of the original variables, thus the sum of the λ_i must equal the sum of the variances (the diagonal elements) of $\{C\}$. Put another way, the traces of $\{\Lambda\}$ and $\{C\}$ must be equal (the trace of a matrix is the sum of the diagonal elements). The correlation matrix $\{R\}$ is the variance-covariance matrix $\{C\}$, where each c_{ij} has been divided by the square root of the product of the i th and j th variances, $(c_{ii}.c_{jj})^{\frac{1}{2}}$. (See section I(iv)) Hence the variance-covariance matrix of the principal components can also be defined as

$$\{\Lambda\} = \{E\}^T.\{R\}.\{E\} \tag{15}$$

where $\{\Lambda\}$ will have different λ_i from the matrix in equation (14), and

$$tr\{\Lambda\} = tr\{R\} = p, \text{ the number of variables} \tag{16}$$

since the diagonal elements of $\{R\}$ are unit.

Reversing equation (15) and postmultiplying by $\{E\}^T$ gives

$$\{E\}^T.\{R\}.\{E\}.\{E\}^T = \{\Lambda\}.\{E\}^T \tag{17}$$

which because $\{E\}.\{E\}^T = \{I\}$ becomes

$$\{E\}^T.\{R\} = \{\Lambda\}.\{E\}^T \tag{18}$$

written out in full

$$\begin{Bmatrix} e_{11} & e_{21} \\ e_{12} & e_{22} \end{Bmatrix} . \begin{Bmatrix} r_{11} & r_{12} \\ r_{21} & r_{22} \end{Bmatrix} = \begin{Bmatrix} \lambda_1 & 0 \\ 0 & \lambda_2 \end{Bmatrix} . \begin{Bmatrix} e_{11} & e_{21} \\ e_{12} & e_{22} \end{Bmatrix}$$

and multiplied out

$$\begin{Bmatrix} e_{11}r_{11}+e_{21}r_{21} & e_{11}r_{12}+e_{21}r_{22} \\ e_{12}r_{11}+e_{22}r_{21} & e_{12}r_{12}+e_{22}r_{22} \end{Bmatrix} = \begin{Bmatrix} \lambda_1 e_{11} & \lambda_1 e_{21} \\ \lambda_2 e_{12} & \lambda_2 e_{22} \end{Bmatrix}$$

subtracting right-hand side from left-hand side, equation (18) becomes

$$\begin{Bmatrix} e_{11}(r_{11}-\lambda_1)+e_{21}r_{21} & e_{11}r_{12}+e_{21}(r_{22}-\lambda_1) \\ e_{12}(r_{11}-\lambda_2)+e_{22}r_{21} & e_{12}r_{12}+e_{22}(r_{22}-\lambda_2) \end{Bmatrix} = \begin{Bmatrix} 0 & 0 \\ 0 & 0 \end{Bmatrix} \quad (19)$$

This gives two sets of simultaneous equations which for a non-zero solution for the e_{ij} to exist must satisfy the condition for each of the λ_i that

$$\begin{vmatrix} r_{11}-\lambda_i & r_{21} \\ r_{12} & r_{22}-\lambda_i \end{vmatrix} = 0$$

or in matrix form ($\{R\}$ is symmetrical, thus $\{R\}=\{R\}^T$)

$$|\{R\}- \lambda_i\{I\}| = 0 \quad (20)$$

Equation (20) is known as the characteristic equation of the matrix $\{R\}$, and the λ_i are the latent roots of $\{R\}$, or its <u>eigenvalues</u>. The matrix $\{E\}$ has as its column vectors the <u>eigenvector</u> associated with each eigenvalue.

The values of the λ_i can be solved from equation (20). Ignoring the subscripts of the λ_i, in the two variable case a quadratic equation is obtained.

$$\begin{vmatrix} r_{11}-\lambda & r_{12} \\ r_{21} & r_{22}-\lambda \end{vmatrix} = (r_{11}-\lambda)(r_{22}-\lambda) - r_{12}r_{21} = 0$$

$$= \lambda^2 - (r_{11}+r_{22})\lambda + (r_{11}r_{22} - r_{12}r_{21}) = 0 \quad (21)$$

The two roots of this quadratic equation are given by

$$\lambda = \frac{-b \pm (b^2 - 4ac)^{\frac{1}{2}}}{2a}$$

where a = 1
       b = $(r_{11}+r_{22})$
       c = $(r_{11}r_{22}-r_{12}r_{21})$

The coefficient b gives the sum of the roots; in this case b=2, which is the number of variables, p, and thus the total variance, to conform to equation (16).

(iii) Principal component loadings

The λ_j can now be substituted back into the simultaneous equations in equation (19) to solve for the e_{ij}. As will be demonstrated in section II(vi), for each value of j and e_{ij} are ratios of each other, and thus can take on an infinite number of related values. They are computed such that $\sum_i e_{ij}^2 = 1$, and the matrix $\{E\}$ becomes the matrix of normalised eigenvectors. This matrix describes the relationship of the principal components to the original variables when both components are variables have unit variance. The components, however, have variances described by $\{\Lambda\}$, thus the product of the matrices $\{E\}$ and $\{\Lambda\}^{\frac{1}{2}}$ gives the weighted relationship of the principal components to the original variables, and is known as the matrix of component loadings, here designated $\{L\}$.

$$\{L\} = \{E\}.\{\Lambda\}^{\frac{1}{2}} \tag{22}$$

The elements of the matrix $\{l_{ij}\}$ are the correlation coefficients between the i th variable and the j th component. Hence they are the cosines of the angles between the i th unit vector and the j th component, and can be visualised as the "projections" of the variables onto the principal components (Gould, 1967, p 79). For the data used in Part I and analysed in section II(vi) these projections are illustrated in fig. 12. The square of any l_{ij} is the proportion of the variance of the i th variable explained the j th component, thus

$$\sum_j l_{ij}^2 = 1 \tag{23}$$

and

$$\sum_i l_{ij}^2 = \lambda_j \tag{24}$$

Equation (23) states that the variance of each variable is preserved; and equation (24) states that it is apportioned according to the λ_j. the variances of the principal components. These follow from the definition of $\{L\}$ as $\{E\}.\{\Lambda\}^{\frac{1}{2}}$, and a little matrix manipulation shows a further interesting property of $\{L\}$.

Postmultiplying $\{L\}$ by its transpose

$$\{L\}.\{L\}^T = \{E\}.\{\Lambda\}^{\frac{1}{2}}.\{\Lambda\}^{\frac{1}{2}T}.\{E\}^T \tag{25}$$

Because $\{\Lambda\}^{\frac{1}{2}}$ is a diagonal matrix $\{\Lambda\}^{\frac{1}{2}}.\{\Lambda\}^{\frac{1}{2}T} = \{\Lambda\}$ so

$$\{L\}.\{L\}^T = \{E\}.\{\Lambda\}.\{E\}^T$$

$\{\Lambda\}$ was earlier defined as $\{E\}^T.\{R\}.\{E\}$, (equation (15)), thus

$$\{L\}.\{L\}^T = \{E\}.\{E\}^T.\{R\}.\{E\}.\{E\}^T = \{I\}.\{R\}.\{I\}$$
$$\{L\}.\{L\}^T \quad \{R\} \tag{26}$$

Premultiplying {L} by its transpose

$$\{L\}^T.\{L\} = \{\Lambda\}^{\frac{1}{2}T}.\{E\}^T.\{E\}.\{\Lambda\}^{\frac{1}{2}} \qquad (27)$$

$$\{L\}^T.\{L\} = \{\Lambda\}^{\frac{1}{2}T}.\{I\}.\{\Lambda\}^{\frac{1}{2}}$$

$$\{L\}^T.\{L\} \quad \{\Lambda\} \qquad (28)$$

Equations (26) and (28) state that the matrix of component loadings {L}, postmultiplied by its transpose produces the correlation matrix {R}, and premultiplied by its transpose produces the diagonal matrix of eigenvalues {Λ}.*

(iv) <u>Principal component scores</u>

The remaining step in this transformation procedure is to relate the original data points, each described by a vector {x}, to the principal components so that each data point may now be described by a vector {y}.

Equation (8) stated that {y} = {A}.{x}. In that equation {x} and {y} were column vectors. If they are transposed to become row vectors, then equation (8) can be rewritten as

$$\{y\} = \{x\}.\{A\}^T \qquad (29)$$

The matrix of eigenvectors {E} was defined as $\{A\}^T$, therefore

$$\{y\} = \{x\}.\{E\} \qquad (30)$$

This equation states that the co-ordinates of any point measured on the component axes are the co-ordinates of that point measured on the variable axes multiplied by the matrix of eigenvectors. These y co-ordinates are known as the <u>component scores</u>. The scores on any one component will have a mean of zero (by definition, as the component axes are transformed from the variable axes by rotating about the multivariate mean, the point 0 in fig. 11), a variance of the eigenvalue, λ_i, of that component (the eigenvalue is a measure of the component length), and will be uncorrelated with the scores on any other component (again, by definition - see section II(ii), equations (13) and (14)).

(v) <u>Further reading</u>

The explanation given above is mathematically far from complete. The only point dwelt upon is the derivation of the characteristic equation (equation 2.18), because this facilitates the solution of the eigenvalues and their associated eigenvectors. Relatively simple treatments of principal components analysis are given in Hope (1968, chapter 4), King (1969, chapter 7), and Rummell (1970, <u>passim</u>, as the book is principally concerned with factor analysis). More advanced, but still comprehensible to the mathematical novice, is Morrison (1967, chapter 7). Hotelling (1933), Thurstone (1947), Anderson (1958), Harman (1960), Lawley and Maxwell (1963), Horst (1965), to select but a few, all provide more complete and essentially mathematical explanations of the technique. The general mathematical problem of rotation (co-ordinate transformation) is covered in texts on group theory, and on classical and quantum mechanics; the author found Goldstein (1950) and Hammermesh (1962) particularly useful.

---

* King (1969, pp 171-2) derives these relations, but his matrix manipulations are in error because he presumably forgot the rule $(\{A\}.\{B\})^T = \{B\}^T.\{A\}^T$

(vi) A worked example

Using the data from Part I, the procedures outlined in sections II(ii), (iii) and (iv) will be worked through. First, the data in table 2 must be converted to standard scores.

	x_1	x_2	x_3
1.	-1.46	-1.10	-0.73
2.	-1.10	-0.73	-0.37
3.	-0.73	-1.46	-1.46
4.	-0.37	+0.37	+0.73
5.	0.00	0.00	0.00
6.	+0.37	-0.37	-1.10
7.	+0.73	+1.10	+1.46
8.	+1.10	+1.46	+1.10
9.	+1.46	+0.73	+0.37

Table 4. Standard scores of data from table 2.

The correlation matrix remains

$$\{R\} = \begin{pmatrix} 1.0000 & 0.8333 & 0.5833 \\ 0.8333 & 1.0000 & 0.9167 \\ 0.5833 & 0.9167 & 1.0000 \end{pmatrix}$$

The eigenvalues can be found by the characteristic equation, equation (20) i.e. $|\{R - \lambda I\}| = 0$.

$$\begin{vmatrix} r_{11}-\lambda & r_{12} & r_{13} \\ r_{21} & r_{22}-\lambda & r_{23} \\ r_{31} & r_{32} & r_{33}-\lambda \end{vmatrix} = 0$$

which gives the cubic equation

$$-\lambda^3 + 3\lambda^2 - (3 - r_{12}^2 - r_{13}^2 - r_{23}^2)\lambda + (1 - r_{12}^2 - r_{13}^2 - r_{23}^2 + 2r_{12}r_{13}r_{23}) = 0$$

substituting in the appropriate values of r_{ij}

$$-\lambda^3 + 3\lambda^2 - 1.1252\lambda + 0.0163 = 0$$

which solves as $\lambda = 2.5636, 0.0151, 0.4214$, and arranging these in descending order

$$\lambda_1 = 2.5636$$
$$\lambda_2 = 0.4214$$
$$\lambda_3 = 0.0151$$
$$\sum_i \lambda_i = 3.0001 \approx 3 = p, \text{ the number of variables}$$

Thus the first principal component, λ_1, accounts for (2.5636/3x100)% of the total variance, i.e. 85.45%; likewise λ_2 accounts for 14.05%; and λ_3 0.5%.

The eigenvectors can be solved from equation (18)

$$\{E\}^T \cdot \{R\} = \{\Lambda\} \cdot \{E\}^T \tag{18}$$

which in this case is

$$\begin{Bmatrix} e_{11} & e_{21} & e_{31} \\ e_{12} & e_{22} & e_{32} \\ e_{13} & e_{23} & e_{33} \end{Bmatrix} \cdot \begin{Bmatrix} r_{11} & r_{12} & r_{13} \\ r_{21} & r_{22} & r_{23} \\ r_{31} & r_{32} & r_{33} \end{Bmatrix} = \begin{Bmatrix} \lambda_1 & 0 & 0 \\ 0 & \lambda_2 & 0 \\ 0 & 0 & \lambda_3 \end{Bmatrix} \cdot \begin{Bmatrix} e_{11} & e_{21} & e_{31} \\ e_{12} & e_{22} & e_{32} \\ e_{13} & e_{23} & e_{33} \end{Bmatrix}$$

which when multiplied out and with right-hand side subtracted from left-hand side yields three sets of simultaneous equations, one set for each of the λ_i. For λ_1 it is

(1) $(r_{11} - \lambda_1)e_{11} + r_{21}e_{21} + r_{31}e_{31} = 0$

(2) $r_{12}e_{11} + (r_{22} - \lambda_1)e_{21} + r_{32}e_{31} = 0$

(3) $r_{13}e_{11} + r_{23}e_{21} + (r_{33} - \lambda_1)e_{31} = 0$

and substituting in the actual values

(1) $-1.5636e_{11} + 0.8333e_{21} + 0.5833e_{31} = 0$

(2) $0.8333e_{11} - 1.5636e_{21} + 0.9167e_{31} = 0$

(3) $0.5833e_{11} + 0.9167e_{21} - 1.5636e_{31} = 0$

Eliminate e_{21} by multiplying (2) by 0.9167/1.5636 = 0.5863

(2) $0.4886e_{11} - 0.9167e_{21} + 0.5375e_{31} = 0$

add (3) $0.5833e_{11} + 0.9167e_{21} - 1.5636e_{31} = 0$

$\overline{1.0719e_{11} \qquad\qquad - 1.0261e_{31} = 0}$

$$1.0719e_{11} = 1.0261e_{31}$$
$$e_{11} = \frac{1.0261}{1.0719}e_{31} = 0.9573e_{31}$$

Eliminate e_{11} by multiplying (2) by $0.5833/0.8333 = 0.7000$

$$\begin{array}{rl}
(2) & 0.5833e_{11} - 1.0945e_{21} + 0.6417e_{31} = 0 \\
\text{subtract } (3) & 0.5833e_{11} + 0.9167e_{21} - 1.5636e_{31} = 0 \\
\hline
& \phantom{0.5833e_{11}} - 2.0112e_{21} + 2.2053e_{31} = 0
\end{array}$$

$$2.0112e_{21} = 2.2053e_{31}$$

$$e_{21} = \frac{2.2053}{2.0112}e_{31} = 1.0965e_{31}$$

These ratios will be obtained no matter which way the equations are solved, so if e_{31} is set equal to 1.0000 the eigenvector becomes

$$\begin{Bmatrix} e_{11} \\ e_{21} \\ e_{31} \end{Bmatrix} = \begin{Bmatrix} 0.9573 \\ 1.0965 \\ 1.0000 \end{Bmatrix}$$

which is normalised by setting the sums of squares of the elements equal to unity, viz.

square the e_{i1} divide by sum of squares   take square root of e_{i1}

$$\begin{Bmatrix} 0.9164 \\ 1.2023 \\ 1.0000 \end{Bmatrix} \qquad \begin{Bmatrix} 0.2938 \\ 0.3855 \\ 0.3206 \end{Bmatrix} \qquad \begin{Bmatrix} 0.5420 \\ 0.6209 \\ 0.5662 \end{Bmatrix}$$

Allowing the computer to do the rest of the work, and to alter the fourth decimal places, the eigenvalues and eigenvectors of the correlation matrix become

$$\{\Lambda\} = \begin{Bmatrix} 2.5636 & 0 & 0 \\ 0 & 0.4214 & 0 \\ 0 & 0 & 0.0150 \end{Bmatrix}$$

$$\{E\} = \begin{bmatrix} 0.5421 & -0.7620 & -0.3542 \\ 0.6209 & 0.0792 & 0.7799 \\ 0.5662 & 0.6427 & -0.5160 \end{bmatrix}$$

Following equation (22), $\{L\} = \{E\}.\{\Lambda\}^{\frac{1}{2}}$

where

$$\{\Lambda\}^{\frac{1}{2}} = \begin{Bmatrix} 1.6011 & 0 & 0 \\ 0 & 0.6492 & 0 \\ 0 & 0 & 0.1225 \end{Bmatrix}$$

then

$$\{L\} = \begin{Bmatrix} 0.8680 & -0.4946 & -0.0434 \\ 0.9914 & 0.0514 & 0.0955 \\ 0.9066 & 0.4172 & -0.0632 \end{Bmatrix}$$

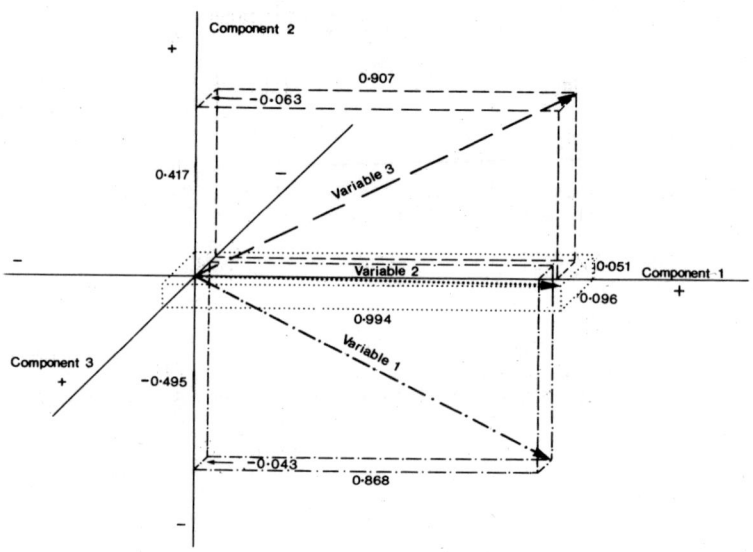

Fig. 12 Component loadings from hypothetical data as "projections" of variables onto principal components

These component loadings are, then, the correlations between each variable and each component, and can be seen as the projections of the unit variable vectors onto the three components, illustrated in fig. 12. Squaring the elements of $\{L\}$ produces the proportion of the variance of each variable "explained" by each component.

$$\{l_{ij}^2\} = \begin{pmatrix} 0.7534 & 0.2446 & 0.0019 \\ 0.9829 & 0.0026 & 0.0091 \\ 0.8219 & 0.1741 & 0.0040 \end{pmatrix}$$

From this matrix it can be seen that variable X_2 is almost completely "explained" by component 1, and variables X_1 and X_3 are 75% and 82% "explained" respectively. Component 2 accounts for most of the rest of the outstanding variance from variables X_1 and X_3, indicating a common but <u>negative</u> association (note the signs in matrix $\{L\}$) between variables X_1 and $\overline{X_3}$ independent of, and much less important than (note the sizes of the first and second eigenvalues) the common trend manifest in component 1. Component 3, accounting for 0.5% of the total variance, simply "mops up" the residue from components 1 and 2.

Given $\{L\}$ and $\{l_{ij}^2\}$, the reader is invited to verify equations (23), (24), (26) and (28).

The computation of the component scores (the values of each data point in units of the principal components) is given by equation (30)

$$\{y\} = \{x\} \cdot \{E\} \tag{30}$$

for data point 1,

$$\{y\} = \{-1.46 \quad -1.10 \quad -0.73\} \cdot \begin{pmatrix} 0.5421 & -0.7620 & -0.3542 \\ 0.6209 & 0.0792 & 0.7799 \\ 0.5662 & 0.6427 & -0.5160 \end{pmatrix}$$

$$= \begin{matrix} -0.7915 & 1.1125 & 0.5171 \\ \{-0.6830 & -0.0871 & -0.8579\} \\ -0.4133 & -0.4692 & +0.3767 \end{matrix}$$

$$= \{-1.8878 \quad 0.5562 \quad 0.0359\}$$

Again, allowing the computer to do the rest of the work, the scores for each data point are

	Component 1	Component 2	Component 3
1.	-1.8855	0.5568	0.0399
2.	-1.2540	0.5422	0.0069
3.	-2.1298	-0.4979	-0.1267
4.	0.4423	0.7765	0.0372
5.	0.0000	0.0000	0.0000
6.	-0.6490	-1.0112	0.0512
7.	1.9031	0.4690	-0.1581
8.	2.1209	-0.0150	0.1859
9.	1.4520	-0.8205	-0.1362
mean	0.0000	-0.0000	+0.0000
sum of squares	20.5088	3.3708	0.1200
N - 1	9-1	9-1	9-1
variance	2.5636	0.4214	0.0150

Table 5. Scores for the nine data points on the three components.

Note that the mean score on each component is zero, and the variance is equal to the eigenvalue. It is also important to realise that the score of a data point on a particular component is simply a linear combination of that point's scores on the original variables. The score of data point 1 on component 1 is

$$y_{11} = 0.5421 x_{11} + 0.6209 x_{12} + 0.5662 x_{13}$$

where the second subscript of the x_{ij} refers to the variables X_1, X_2, X_3.

All that is being done in principal components analysis is the transformation of the system of reference from one set of axes (the variables) to another (the principal components). Nothing "new" is created; nothing "old" is lost.

(vii) The significance of principal components

Unless there is perfect correlation between two or more of the variables, then p principal components are required to account for the p-dimensional variable space. Hence <u>all</u> principal components are significant. This can be verified by considering that the product of the eigenvalues of a matrix is equal to the determinant of that matrix. If the determinant is non-zero then none of the eigenvalues (the dimensions of the principal components) can be zero.

When two variables have perfect positive intercorrelation, their correlations with every other variable will be identical; thus two rows (and two columns) of the matrix will be the same. When two variables have perfect negative intercorrelation, then their correlations with each other variable will differ only in sign.

In both cases there is a row (or column) which is a simple linear combination of another row (or column); in the first case the linear combination is +1, in the second case -1. When one row (or column) of a matrix is a simple linear combination of another, the determinant will always equal zero. Thus at least one of the latent roots (eigenvalues) must equal zero, and the matrix contains "dependency" or "redundancy". The determinant will also be zero if one row (or column) contains all zero elements, but this is never the case with a simple correlation matrix, because the correlation of a variable with itself is unity.

The null hypothesis, then, that $\lambda_j = 0$ must be rejected except in these unusual circumstances. A very useful hypothesis to test, however, is that at any stage of the extraction of the eigenvalues

$$H_o: \lambda_i = \lambda_{i+1} = \cdots\cdots\cdots\cdots\cdots = \lambda_p$$

The hypothesis that the roots do not differ from each other infers, for example in the three variable case that the probability field is spherical, and thus there is an infinite number of positions that three equal and orthogonal axes can assume. Hence there is an infinite suite of eigenvectors, meaning that the relationship between the principal components and the original variables has an infinity of definitions. If this null hypothesis cannot be rejected, then further extraction of eigenvalues is pointless.

The determinant of a matrix is a measure of the volume of the ellipsoid defined by the elements of the matrix (Gould, 1967, pp 56-57; Hope, 1968, p.65). If the total length of the orthogonal axes which define the ellipsoid remains unchanged, then the volume of the ellipsoid increases as the lengths of the axes become more similar, and reaches a maximum when the axes of the ellipsoid are of equal length, i.e. the ellipsoid has become spherical.

If t is the number of eigenvalues (axes) under consideration, then the length of the axes when all are equal is given as

$$\left(\frac{1}{t} \cdot \sum_{i=1}^{t} \lambda_i \right)$$

and thus the volume of this sphere is

$$\left(\frac{1}{t} \cdot \sum_{i=1}^{t} \lambda_i\right)^t$$

So the ratio, Q, of the actual volume to the maximum possible volume is

$$Q = \frac{|R|}{\left(\frac{1}{t} \sum_{i=1}^{t} \lambda_i\right)^t} \qquad (31)$$

and testing H_o: $\lambda_i = \lambda_{i+1} = \ldots\ldots\ldots = \lambda_p$ becomes a matter of testing whether Q is significantly less than unity. Bartlett (1950; 1951a & b) has established that

$$\chi^2 = -(N - p - \tfrac{1}{2}) \cdot \log_e Q \qquad (32)$$

where N is the sample size and p is the total number of eigenvalues (i.e. the total number of variables) with degrees of freedom

$$d.f. = \tfrac{1}{2}(p - k + 2)(p - k - 1) \qquad (33)$$

where k is the number of eigenvalues not under consideration (k=p-t). Q can also be computed as q^t, where q is the ratio of the geometric mean to the arithmetic mean of the t eigenvalues.

$$q = \frac{\left(\prod_{i=1}^{t} \lambda_i\right)^{\frac{1}{t}}}{\left(\frac{1}{t} \cdot \sum_{i=1}^{t} \lambda_i\right)} \qquad (34)$$

Thus for the data analysed below

$$q = \frac{(2.5636 \times 0.4214 \times 0.0150)^{1/3}}{\tfrac{1}{3}(2.5636 + 0.4214 + 0.0150)} = \frac{(0.0162)^{1/3}}{\tfrac{1}{3}(3)} = 0.253$$

$$Q = q^t = 0.253^3 = 0.0162$$

$$\chi^2 = -(9 - 3 - \tfrac{1}{2}) \cdot \log_e 0.0162$$
$$\chi^2 = -(5\tfrac{1}{2}) \cdot (-4.1228)$$
$$\chi^2 = 22.6754$$

with degrees of freedom

$$d.f. = \tfrac{1}{2}(3 - 0 + 2) \cdot (3 - 0 + 1)$$
$$= \tfrac{1}{2}(5) \cdot (2)$$
$$= 5$$

The critical value of χ^2 with 5 d.f., at $\alpha=0.05$, is 11.07; so the null hypothesis that the eigenvalues are equal can be rejected with 95% confidence. The next step is the focus attention on the last p - 1 roots, in this case to test the null hypothesis that λ_2 and λ_3 are equal. Hence

$$q = \frac{(0.4214 \times 0.0150)^{\frac{1}{2}}}{\frac{1}{2}(0.4214 + 0.0150)} = \frac{0.0794}{0.2182} = 0.3639$$

$$Q = q^2 = 0.1324$$

$$\chi^2 = -(9 - 3 - \tfrac{1}{2}).\log_e 0.1324$$

$$\chi^2 = -5\tfrac{1}{2} \times 2.0220$$

$$\chi^2 = 11.121$$

$$\text{d.f.} = \tfrac{1}{2}(3 - 1 + 2).(3 - 1 - 1)$$

$$\text{d.f.} = 2$$

The critical value of chi-square with 2 d.f. at $\alpha=0.05$ is 5.99; so the null hypothesis that $\lambda_2=\lambda_3$ may be rejected with 95% confidence.

Further, the difference between the chi-square values for three and for two eigenvalues, i.e. 22.6754 - 11.1210 = 11.5544 with 5 - 2 = 3 d.f., can be used to test the null hypothesis that the difference between λ_1 and λ_2 was not contributing to the overall heterogeneity established between λ_1, λ_2 and λ_3. The critical chi-square value for $\alpha=0.05$ and 3 d.f. is 7.81, so the null hypothesis may be rejected with 95% confidence.

These tests apply only when the data are multivariate normally distributed, and strictly speaking only to the eigenvalues derived from a variance-covariance matrix. When applied to the eigenvalues from a correlation matrix, the test is only approximate. An exact test for the last p-1 eigenvalues of a correlation matrix is given in section IV(vi).

III THE TECHNIQUE APPLIED

(i) Introduction

In a study of the changes in the agricultural structure of the Irish Republic since its inception as the Irish Free State in 1922, Walsh (1974) collected data at a county level pertaining to arable, pastural and other land uses, types of stock, sizes of holdings, degree of mechanisation, and size of the agricultural labour force. In one part of the study he derived 39 variables expressing percentage changes of these features from 1951 to 1971. In order to examine the structure of the correlations between these variables, and subsets of these variables, he applied principal components analysis, and then used some of the results to classify the Irish Republic into regions of agricultural change, using a hierarchical grouping procedure on the component scores.

Fig. 13 Irish counties

The study which follows uses Walsh's data; it does not repeat his selection of variables, and thus produces slightly different results - a point which will be pursued later.

(ii) <u>The data</u>.

For the 26 counties of the Irish Republic (fig. 13), 15 variables describing percentage change 1951-71 were selected:

County	(i)	(ii)	(iii)	(iv)	(v)	(vi)	(vii)	(viii)	(ix)	(x)	(xi)	(xii)	(xiii)	(xiv)	(xv)
1. Carlow	12.6	-9.9	-2.8	11.5	-6.0	-10.0	-92.0	67.8	17.2	124.1	-5.6	11.0	8.0	15.4	-36.0
2. Dublin	88.1	-8.2	-19.7	-27.5	35.5	-36.4	-88.2	-40.9	30.1	78.2	-7.0	11.1	5.3	18.5	-62.7
3. Kildare	39.1	-36.7	-0.5	13.7	-20.2	-28.4	-88.2	85.0	35.8	207.7	-1.0	18.3	5.8	17.5	-39.0
4. Kilkenny	5.1	-36.6	0.2	34.2	-25.5	-36.6	-80.7	54.4	4.2	142.7	-2.9	6.4	5.9	23.0	-37.4
5. Laois	-17.7	-34.9	9.9	16.9	-15.6	-54.3	-77.2	82.0	10.7	149.5	-8.4	24.2	9.3	15.8	-41.2
6. Longford	-77.5	-74.9	8.1	15.1	-5.4	-75.4	-34.8	80.0	21.2	95.2	-15.2	39.7	4.2	7.7	-49.2
7. Louth	30.0	-64.3	-7.1	26.9	0.4	-77.0	-77.0	46.5	5.1	125.4	-4.3	24.6	7.6	20.3	-45.3
8. Meath	50.4	-44.7	-5.8	21.0	-19.7	7.6	-84.3	84.3	35.5	126.9	-4.3	28.5	5.3	21.4	-36.2
9. Offaly	-21.5	-25.4	14.7	11.6	-17.1	-55.0	-76.4	74.0	3.7	128.4	-7.7	16.7	7.1	11.7	-39.5
10. Westmeath	0.9	-52.2	2.5	27.7	-17.1	-26.2	-77.8	93.6	35.9	76.6	8.4	29.5	4.1	12.1	-38.7
11. Wexford	1.9	-10.8	-8.5	50.4	-4.7	-10.9	-84.0	97.5	20.9	136.4	-4.3	5.9	9.1	19.1	-38.4
12. Wicklow	43.0	-29.1	-1.3	10.6	-4.5	-10.6	-85.6	47.4	2.5	137.6	15.8	3.1	15.7	-42.1	
13. Clare	-74.0	-29.5	14.2	15.8	-18.6	-75.0	-19.5	45.7	5.2	5.5	-14.3	19.2	1.6	11.3	-37.7
14. Cork	12.7	-29.5	-2.3	25.3	-6.5	-66.2	-80.5	46.1	1.2	-14.3	0.9	7.1	6.1	31.4	-41.6
15. Kerry	-48.9	-57.1	2.2	15.6	-9.6	-87.7	-52.6	25.9	3.1	35.9	-0.1	8.0	5.1	27.5	-39.9
16. Limerick	-78.8	-78.1	2.6	19.4	-8.7	-66.8	-56.5	32.6	1.7	1.9	-6.0	10.3	8.8	23.4	-42.1
17. Tipperary	-17.3	-48.9	3.8	19.0	-11.4	-49.5	-79.1	57.0	6.9	85.3	-6.0	14.9	5.8	22.8	-41.6
18. Waterford	12.6	-47.6	1.4	92.2	-17.9	-58.7	-57.7	53.9	6.6	55.9	-9.3	6.2	4.4	22.4	-39.2
19. Galway	-44.3	-40.6	11.9	22.8	-6.1	-83.1	-26.5	64.4	-1.5	45.6	-13.3	36.7	7.0	-36.7	
20. Leitrim	-86.6	-75.0	6.0	13.7	-5.1	-100.0	44.3	18.6	-0.3	90.2	-20.1	80.7	5.1	-41.4	
21. Mayo	-62.9	-66.3	17.3	16.0	-4.7	-89.7	-12.4	39.4	-1.1	82.1	-19.0	64.5	0.6	7.7	-43.4
22. Roscommon	-75.4	-63.4	5.8	15.4	-5.3	-92.5	-26.8	45.4	-0.2	83.7	-28.0	75.4	2.5	5.1	-40.3
23. Sligo	-75.2	-73.4	9.9	6.9	-2.6	-73.2	-11.5	29.9	-3.5	59.6	-24.4	70.9	7.2	11.1	-43.8
24. Cavan	-86.4	-73.4	3.4	31.2	10.5	81.0	-31.6	60.1	2.4	100.6	-14.3	59.1	7.2	18.5	-46.5
25. Donegal	-35.2	-48.0	22.8	22.0	-3.3	-70.4	-46.6	15.4	-0.4	53.1	-2.6	17.5	3.2	8.2	-50.9
26. Monaghan	-73.6	-74.4	14.9	44.6	-8.6	56.6	-40.0	64.1	2.3	123.8	-15.0	51.2	5.3	21.8	-44.8

Table 6. Changes, 1951 - 1971, by county, for 15 agricultural variables, Republic of Ireland.

Table 7. Lower triangle of correlation matrix for 15 variables, first analysis.

	(i)	(ii)	(iii)	(iv)	(v)	(vi)	(vii)	(viii)	(ix)	(x)	(xi)	(xii)	(xiii)	(xiv)	(xv)
(i) "CORN"	1.00														
(ii) "ROOT"	0.70	1.00													
(iii) "PASTURE"	-0.71	-0.41	1.00												
(iv) "HAY"	-0.08	-0.15	0.11	1.00											
(v) "OTHER"	0.10	0.04	-0.37	-0.46	1.00										
(vi) "WHEAT"	0.27	0.05	-0.32	0.21	0.15	1.00									
(vii) "OATS"	-0.80	-0.64	0.55	-0.02	0.09	-0.40	1.00								
(viii) "MILCH"	0.00	0.06	0.12	0.45	-0.66	0.31	-0.29	1.00							
(ix) "P-MILCH"	0.60	0.38	-0.54	-0.14	-0.02	0.28	-0.53	0.34	1.00						
(x) "SHEEP"	0.43	0.25	-0.24	-0.00	-0.16	0.49	-0.41	0.47	0.42	1.00					
(xi) "1-10 HO"	0.66	0.46	-0.38	0.03	-0.16	0.27	-0.73	0.17	0.37	0.21	1.00				
(xii) "50-100"	-0.64	-0.67	0.40	-0.13	0.17	-0.09	0.80	-0.08	-0.29	-0.04	-0.78	1.00			
(xiii) "COMBINE"	0.42	0.36	-0.46	0.12	0.00	0.54	-0.77	0.34	0.25	0.41	0.51	-0.61	1.00		
(xiv) "MILKMAC"	0.41	0.18	-0.51	0.28	-0.08	0.38	-0.58	0.01	0.09	0.00	0.49	-0.61	0.60	1.00	
(xv) "MALES"	-0.15	0.07	0.20	0.41	-0.84	-0.12	-0.01	0.70	0.00	0.10	0.07	-0.07	0.01	0.02	1.00

    (i) proportion of land under corn crops - "CORN";
   (ii) proportion of land under root and bean crops - "ROOT";
  (iii) proportion of land under pasture - "PASTURE";
   (iv) proportion of land under hay - "HAY";
    (v) proportion of land under other uses - "OTHER";
   (vi) land under wheat as a proportion of land under corn crops - "WHEAT";
  (vii) land under oats as a proportion of land under corn crops - "OATS";
 (viii) the number of milch cows - "MILCH";
   (ix) milch cows as a proportion of all cattle - "P-MILCH";
    (x) the number of sheep - "SHEEP";
   (xi) the number of holdings between 1 and 10 acres as a proportion of all holdings - "1-10 HO";
  (xii) the number of holdings between 50 and 100 acres as a proportion of all holdings - "50-100";
 (xiii) the number of combine harvesters per 1000 acres of corn - "COMBINE";
  (xiv) the number of milking machines per 1000 milch cows - "MILKMAC";
   (xv) the number of males employed in agriculture - "MALES".

The data are given in table 6.

(iii) <u>The first analysis</u>

The 15 x 15 matrix of correlations was computed from the data, and its lower triangle is given in table 7. The null hypothesis that any r_{ij} was equal to zero was tested by computing the ratio of the explained variance to the unexplained variance, and comparing this to the critical values of Snedecor's F for degrees of freedom 1 and N-2. For $\alpha=0.05$, and d.f. 1 and 24, an r_{ij} must equal or exceed 0.39 for the null hypothesis to be rejected. For 42 of the 105 r_{ij} the null hypothesis could be rejected with 95% confidence.

The eigenvalues and eigenvectors of the correlation matrix were derived, and the eigenvectors scaled by the square root of the corresponding eigenvalue to produce the matrix of component loadings. The eigenvalues were tested for their heterogeneity by the Bartlett tests. The results for the first five principal components are given in table 8.

Converting the data matrix to a matrix of standard scores and post-multiplying it by the matrix of eigenvectors gives the matrix of component scores. The scores on the first two components for each county are given in table 9, and are shown mapped in figures 14 and 15.

The first principal component accounts for 38% of the total variance; the second a further 20%; the third a further 11%; and the fourth 10%; making 79% of the total variance "explained" by four uncorrelated combinations of the original variables. The components can be interpreted in terms of the variables which load "most heavily" onto them (i.e. have the highest component loadings).

The first component has high positive loadings from corn crops, root crops, proportion of milch cows, 1-10 acre holdings, combine harvesters and milking machines, and high negative loadings from pasture, oats and 50-100 acre holdings. This component then, is describing the general trend of correlations resulting from areas with:
- highest rates of increase in corn crops, proportion of milch cows, combine harvesters and milking machines;
- lowest rates of increase in pasture and 50-100 acre holdings;

	Variable	I	II	Component III	IV	V
(i)	"CORN"	0.850	0.258	0.026	-0.245	0.150
(ii)	"ROOT"	0.669	0.134	-0.145	-0.476	-0.150
(iii)	"PASTURE"	-0.682	-0.411	-0.104	-0.071	-0.458
(iv)	"HAY"	0.077	-0.635	-0.177	0.449	0.305
(v)	"OTHER"	-0.066	0.901	0.190	0.209	-0.020
(vi)	"WHEAT"	0.484	-0.046	0.508	0.580	-0.077
(vii)	"OATS"	-0.948	-0.000	0.056	0.033	0.135
(viii)	"MILCH"	0.295	-0.837	0.338	-0.046	-0.024
(ix)	"P-MILCH"	0.600	0.038	0.423	-0.363	0.400
(x)	"SHEEP"	0.477	-0.213	0.697	-0.088	-0.211
(xi)	"1-10 HO"	0.783	-0.034	-0.270	-0.074	-0.126
(xii)	"50-100"	-0.800	0.013	0.525	0.077	0.122
(xiii)	"COMBINE"	0.772	-0.072	0.038	0.366	-0.320
(xiv)	"MILKMAC"	0.631	-0.008	-0.384	0.522	0.175
(xv)	"MALES"	0.041	-0.872	-0.125	-0.235	0.121
Eigenvalues		5.703	2.983	1.655	1.489	0.752
Variance explained		38.0%	19.9%	11.0%	10.0%	5.0%
Chi-square (1)		168.8	126.9	100.0	84.6	64.3
d.f. (1)		119	104	90	77	65
Chi-square (2)		41.9	26.9	15.4	20.3	8.3
d.f. (2)		15	14	13	12	11

Table 8. Component loadings, eigenvalues, and chi-square values, first analysis.

- highest rates of decrease in oats;
- lowest rates of decrease in root crops and 1-10 acre holdings.

The interpretation is complex because the data can be positive or negative. All that a positive loading means is that there is positive correlation between the component and that variable. The number of milking machines per 1000 milch cows has increased in every county, i.e. all the data are positive, and the positive loading indicates that the areas with the highest increases score high on that component. However, the proportion of land under root and bean crops has declined in every county, i.e. all the data are negative; so the positive loading between component 1 and this variable indicates that counties with the highest values on the continuum from $-\infty$ to $+\infty$ (in fact, the smallest decreases) score high on that component.

The counties with this overall trend, then, score high on component 1, and the map of the scores (fig. 14) shows these counties to be those of east Leinster, in particular Wexford, Kildare, Wicklow, Meath, Dublin, and Carlow.

But contributing equally to component 1 are the areas with precisely the opposite characteristics: highest rates of increase in pasture and 50-100 acre holdings; lowest rates of increase in milking machines; lowest rates of increase and some decreases in combine harvesters and proportion of milch cows; highest rates of decrease in corn crops, root crops, and 1-10 acre holdings; lowest rates of decrease in oats.

	County	First analysis		Second analysis		
		I	II	I	II	III
1.	Carlow	2.52	-0.26	2.83	-0.31	-0.35
2.	Dublin	2.56	7.34			
3.	Kildare	2.86	-1.17	3.51	-1.21	-1.77
4.	Kilkenny	1.81	-1.43	2.29	-0.87	0.83
5.	Laois	1.17	-1.12	1.53	-0.37	-0.67
6.	Longford	-1.77	-0.24	-1.63	-0.21	-1.61
7.	Louth	1.61	1.21	1.03	3.12	-0.31
8.	Meath	2.62	-1.18	3.35	-1.02	-1.40
9.	Offaly	0.54	-1.14	0.98	-1.14	-0.02
10.	Westmeath	1.45	-1.35	2.08	-1.38	-1.05
11.	Wexford	3.21	-1.13	3.51	0.22	-0.94
12.	Wicklow	2.74	0.60	2.54	1.25	0.83
13.	Clare	-2.23	-0.70	-1.77	-2.42	1.41
14.	Cork	1.72	0.47	1.56	0.94	2.50
15.	Kerry	0.01	0.33	-0.13	0.38	2.50
16.	Limerick	-0.43	0.24	-0.68	1.35	2.12
17.	Tipperary	0.80	-0.22	0.87	0.34	0.82
18.	Waterford	0.63	-1.76	0.85	-0.70	1.37
19.	Galway	-2.13	-0.69	-1.71	-1.78	0.18
20.	Leitrim	-4.90	0.57	-4.85	-0.97	-0.70
21.	Mayo	-3.76	0.18	-3.77	-0.70	-0.42
22.	Roscommon	-3.56	0.05	-3.33	-0.95	-1.05
23.	Sligo	-3.84	0.80	-3.97	0.19	-0.30
24.	Cavan	-0.93	0.51	-1.51	3.61	-2.02
25.	Donegal	-1.66	1.18	-2.37	0.92	1.50
26.	Monaghan	-1.06	-1.09	-1.19	1.73	-1.54

Table 9. Scores on first two components, first analysis; and first three components, second analysis.

These counties have high negative scores on component 1, the map (fig. 14) showing the major area to be in Connacht, in particular counties Leitrim, Sligo, Mayo and Roscommon.

Counties which have scores near zero (Offaly, Kerry, Limerick, Waterford) will be exhibiting either near mean values for these variables, or will have some characteristics of the area scoring positively on the component, and some characteristics of the area scoring negatively.

The first component, accounting for 38% of the total variance, can be interpreted as an "intensification of agriculture" component, showing the positive aspects of this trend to have been most marked in east Leinster, the principal agricultural region of the country, and to be characterised by an increasing emphasis on corn crops and dairying, with much mechanisation, root crops not declining in importance as much as in other areas, a very low increase in pasture, and a very heavy decline in the already small amount of oats. In contrast, the west, Connacht in particular, has tended to move towards pasture, with a concomitant lower emphasis on mechanisation. The variables relating to size of holdings have an interesting relation to this component. Increases in 50-100 acre holdings have been lowest in the "intensive" east, highest in the less intensive west; decreases in 1-10 acre holdings have been least in the east, most in the west. This reflects the fact that

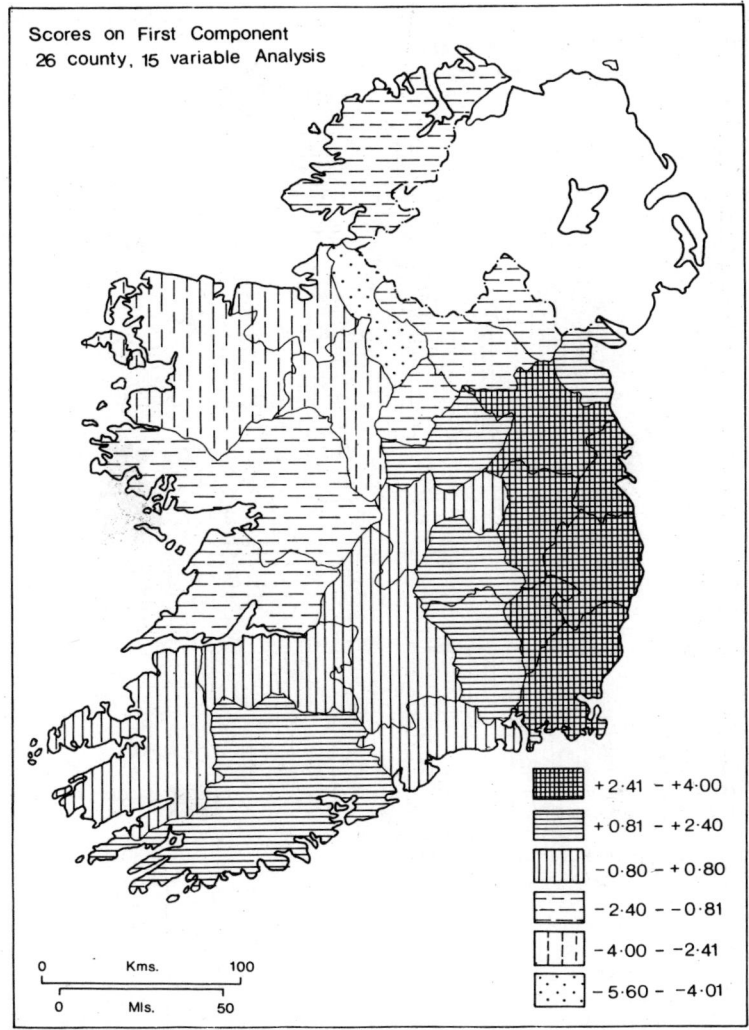

Fig. 14 First Analysis: Component I

landholding structure in the east of the country has been stable for most of this century, but that in the west there is a massive consolidation in progress at present, with the creation of larger, more viable holdings. Agriculture is becoming more progressive in all parts of the Irish Republic, but the nature of the progress shows marked east-west contrasts.

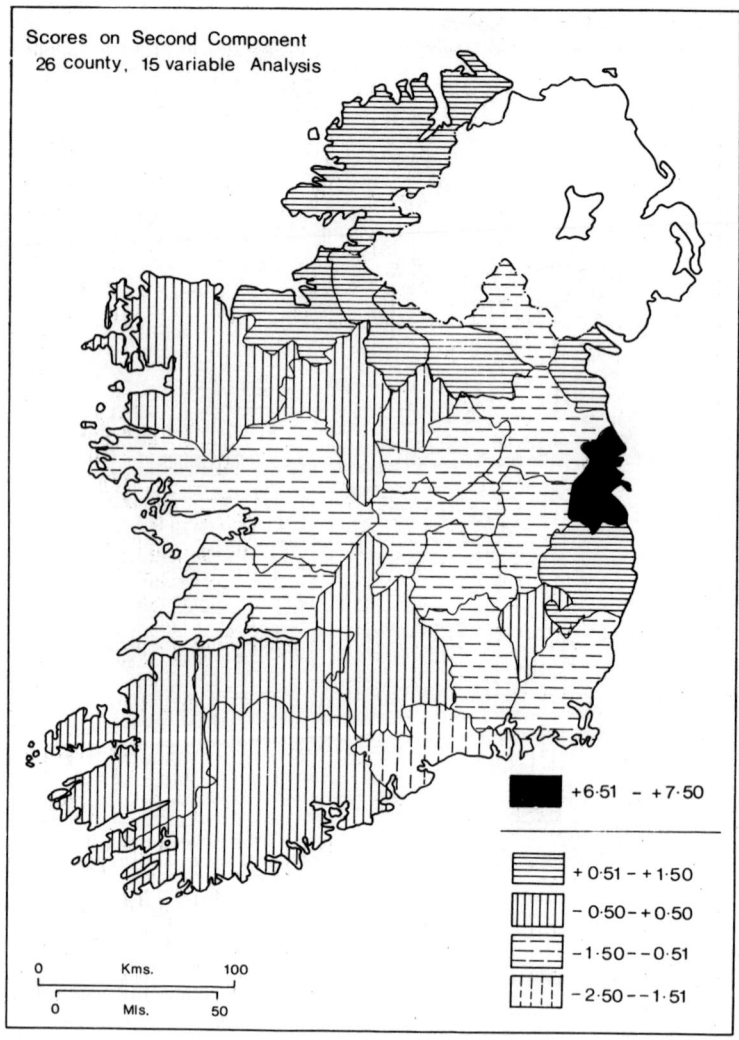

Fig. 15 First Analysis : Component II

The second component, accounting for 20% of the total variance, has a high positive loading from other land, and high negative loadings from hay, milch cows, and males in agriculture. This component is describing the trends in correlation resulting from areas with the highest rates of increase in other land, and highest rates of decrease in hay, milch cows, and males in agriculture. The map of component scores (fig. 15) makes it clear that this area is Co. Dublin, the only county with a decrease in the proportion of land under hay and in the number of milch cows, by far the largest in other land, and the greatest decline in the agricultural labour force. This component is describing a unique area in the Irish Republic - the only large built-up area in the country - and at a time when its development has been most rapid. So anomalous is Co. Dublin that the scores of the other counties are meaningless, except perhaps that of Louth, the smallest county, which has two rapidly developing medium-sized (by Irish standards) towns, Drogheda and Dundalk.

The chi-square values for testing the null hypothesis that the eigenvalues are all equal are given in table 8. Working at the 95% level the null hypothesis that all 15 eigenvalues are equal can be rejected; so too can the null hypothesis that the last 14 are equal. The hypotheses that λ_1 and λ_2 are not contributing to these heterogeneities can also be rejected. But the hypothesis that $\lambda_3 = \lambda_4 = \ldots = \lambda_{15}$ cannot be rejected with 95% confidence, thus there is little point in trying to interpret these components.

(iv) The second analysis

Given that the second largest component of the first analysis referred to Co. Dublin and very little else, it was decided to exclude the Co. Dublin data, and reanalyse the remaining 25 counties. The first result of this was to change all the correlation coefficients, for N=25, the r_{ij} must equal or exceed 0.40 for the null hypothesis that the r_{ij}=0 to be rejected with 95% confidence, and for 48 of the 105 r_{ij} this could be done. However, the variable pertaining to hay (no. iv) was not significantly correlated with any other variable, and thus was excluded, leaving 14 variables measured for each of 25 counties. The same procedure as in the first analysis was followed, and the results are set out in tables 9, 10 and 11.

The result of the changes was to "strengthen" the first component, which accounted for 45% of the total variance. This was in part due to the omission of the hay variable, reducing the total variance from 15 to 14, and in part due to the removal of the Dublin anomaly, which increased the component loadings relating to milch cows and sheep in particular. This first component can be interpreted in virtually the same way as in the first analysis, having the same pattern of component loadings (fig. 16).

The second and third components, respectively accounting for 15% and 13% of the total variance, relate to specific areas of the country which produce trends in correlation independent of the major trend described by the first component. The second component has high positive loadings from other land and wheat, and a high negative loading from males in agriculture. The counties which score high on this component have: increases in wheat as a proportion of corn crops, as opposed to the national trend for wheat acreage to decline in favour of feeding barley; increases, or very small rates of decrease, in other land; and the highest rates of decrease in males employed in agriculture. It is the border counties of Cavan, Louth and Monaghan which most strongly exhibit this pattern, with the western counties of Clare and Galway most strongly exhibiting the converse (fig. 17). Certainly the border counties are the only ones to show increases in the proportion of corn acreage under wheat, whilst the western counties have shown the largest decreases; and the overall decline of the agricultural labour force has been most marked in the border counties, least marked (though still large) in the east and in Galway

		(i)	(ii)	(iii)	(v)	(vi)	(vii)	(viii)	(ix)	(x)	(xi)	(xii)	(xiii)	(xiv)	(xv)
(i)	"CORN"	1.00													
(ii)	"ROOT"	0.64	1.00												
(iii)	"PASTURE"	-0.62	-0.28	1.00											
(v)	"OTHER"	-0.40	-0.38	0.04	1.00										
(vi)	"WHEAT"	0.28	0.04	-0.35	0.20	1.00									
(vii)	"OATS"	-0.81	-0.62	0.54	0.37	-0.40	1.00								
(viii)	"MILCH"	0.43	0.43	-0.34	-0.34	0.43	-0.56	1.00							
(ix)	"P-MILCH"	0.53	0.29	-0.34	-0.43	0.28	-0.50	0.78	1.00						
(x)	"SHEEP"	0.52	0.29	-0.45	-0.18	0.49	-0.44	0.57	0.47	1.00					
(xi)	"1-10 HO"	0.73	0.49	-0.32	-0.43	0.26	-0.74	0.24	0.39	0.21	1.00				
(xii)	"50-100"	-0.65	-0.66	0.38	-0.27	-0.09	0.79	-0.24	-0.25	-0.05	-0.79	1.00			
(xiii)	"COMBINE"	0.47	0.38	-0.44	0.43	0.26	-0.79	0.46	0.25	0.41	0.52	-0.61	1.00		
(xiv)	"MILKMAC"	0.42	0.17	-0.53	-0.02	0.54	-0.58	0.07	0.08	0.01	0.49	-0.60	0.60	1.00	
(xv)	"MALES"	0.30	0.56	-0.33	-0.63	-0.15	-0.24	0.43	0.40	0.08	0.12	-0.27	0.03	0.10	1.00

Table 10. Lower triangle of correlation matrix for 14 variables, second analysis.

(i)	"CORN"	0.866	-0.044	0.050	0.140	0.217
(ii)	"ROOT"	0.686	-0.352	0.185	0.321	-0.413
(iii)	"PASTURE"	-0.672	-0.215	0.053	0.516	-0.028
(v)	"OTHER"	-0.456	0.687	-0.152	0.072	-0.209
(vi)	"WHEAT"	0.440	0.583	-0.472	-0.059	-0.086
(vii)	"OATS"	-0.932	-0.125	-0.101	-0.140	0.009
(viii)	"MILCH"	0.658	-0.228	-0.580	-0.022	-0.137
(ix)	"P-MILCH"	0.641	-0.307	-0.471	-0.133	0.351
(x)	"SHEEP"	0.524	0.047	-0.644	0.263	0.007
(xi)	"1-10 HO"	0.762	0.146	0.325	0.223	0.318
(xii)	"50-100"	-0.774	-0.008	-0.550	-0.156	0.020
(xiii)	"COMBINE"	0.737	0.469	0.003	0.022	-0.293
(xiv)	"MILKMAC"	0.580	0.417	0.411	-0.427	0.013
(xv)	"MALES"	0.429	-0.716	0.037	-0.381	-0.299

Eigenvalues	6.305	2.076	1.847	0.908	0.689
Variance explained	45.0%	14.9%	13.2%	6.5%	4.9%
Chi-square (1)	151.1	98.8	82.5	60.9	52.2
d.f. (1)	104	90	77	65	54
Chi-square (2)	52.3	16.3	21.6	8.7	6.8
d.f. (2)	14	13	12	11	10

Table 11. Component loadings, eigenvalues, and chi-square values, second analysis

and Clare. The category "other land" covers a multitude of land uses, and its precise meaning in this context is not clear. Generally, the second component seems to be describing separate trends which happen to be occuring in the same areas.

The third component is somewhat easier to interpret. The highest loadings are all negative, and are from the numbers of milch cows, the numbers of sheep, and the proportion of 50-100 acre holdings. Thus the areas which score high on this component will have the lowest rates of increase in milch cows, sheep and 50-100 acre holdings. These turn out to be Kerry, Cork and Limerick, where the dairying industry was already well-established by the 1950's, and thus the rapid increase shown over the rest of the country is not repeated there (fig. 18). It would seem that the pattern of increase in sheep numbers coincides with the dairying pattern, rather than being part of the same process.

A major factor in the lack of clarity of the second and third components becomes immediately apparent when the chi-square values from the Bartlett tests are examined. the null hypothesis that $\lambda_1 = \lambda_2 = \ldots\ldots\ldots = \lambda_{14}$ can be rejected with 95% confidence; so too can the null hypothesis that λ_1 is not contributing to the heterogeneity. But the null hypotheses concerning the remaining 13 eigenvalues cannot be rejected with 95% confidence. Thus after the first principal component has accounted for 45% of the total variance, the 13-dimensional ellipsoid defining the remaining 55% is not significantly different from a 13-dimensional sphere, meaning that the values of the

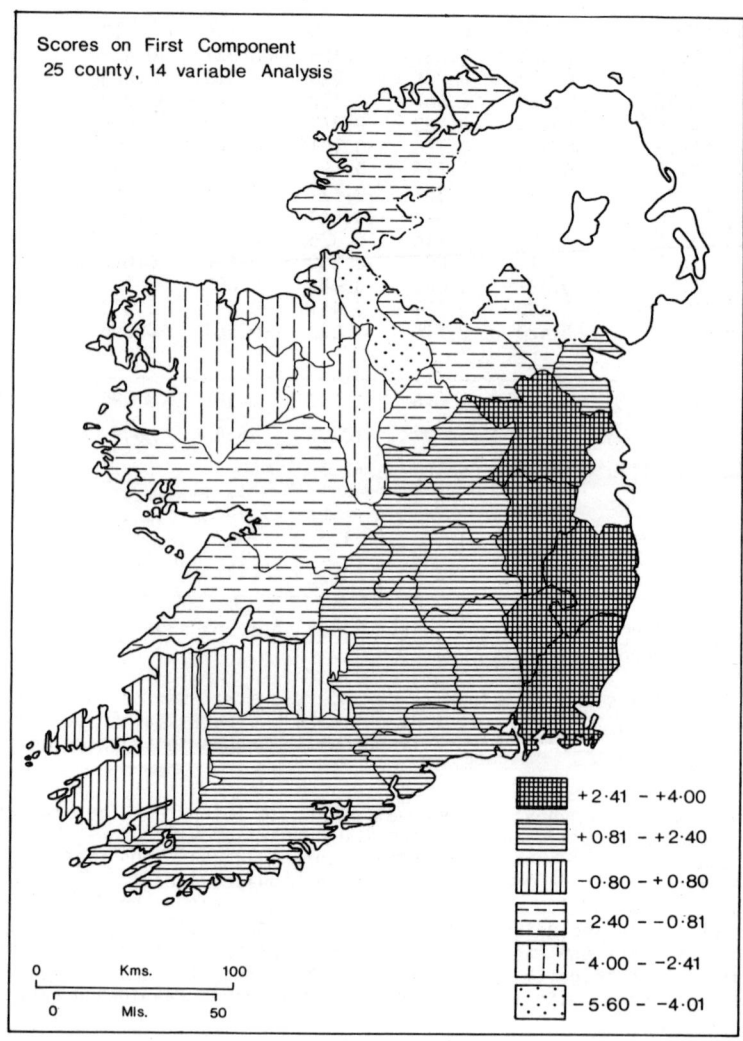

Fig. 16 Second Analysis: Component I

eigenvectors can be seen as largely a matter of chance. Discussion of the second and third components, then, has no statistical basis.

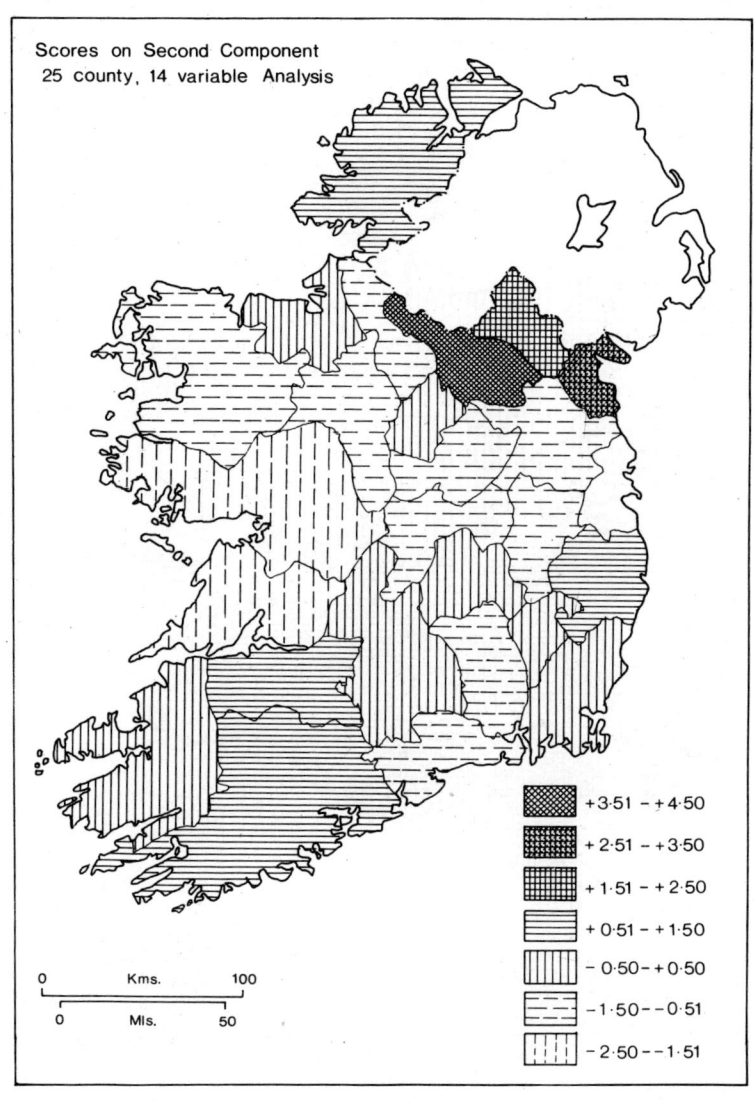

Fig. 17   Second Analysis : Component II

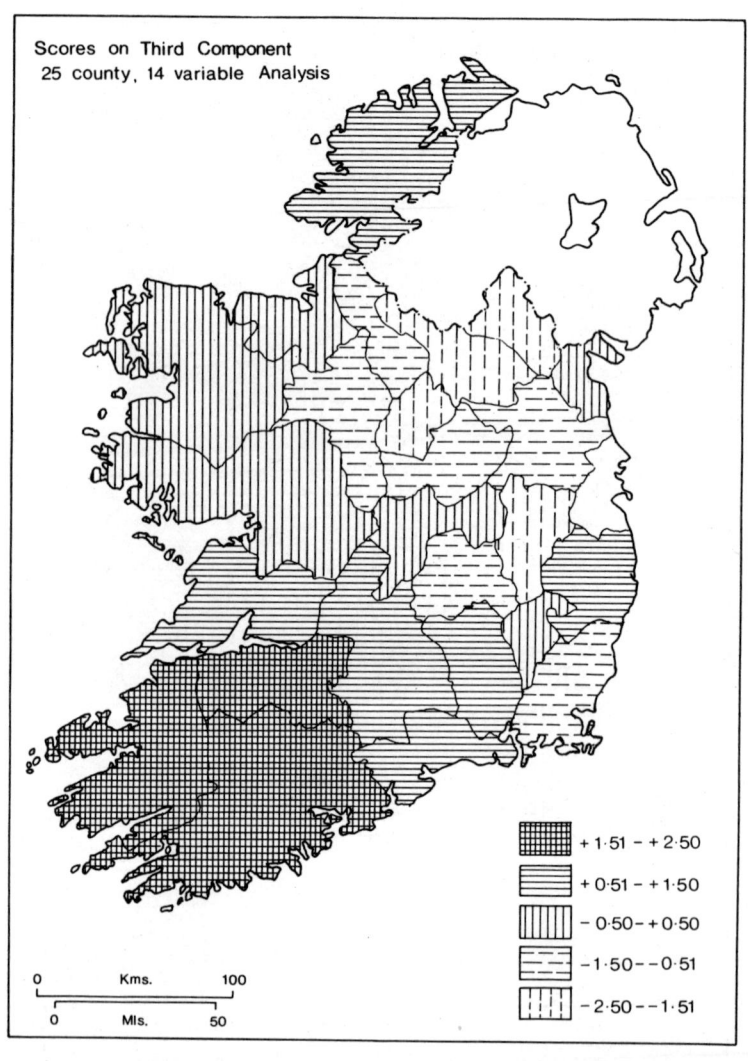

Fig. 18 Second Analysis : Component III

IV THE TECHNIQUE ELABORATED

(i) The normal distribution and the principal components of correlation matrices

Most principal components analyses published by geographers have been analyses of correlation matrices where the measure of correlation used is a sample estimate, r, of Pearson's product moment correlation coefficient, ρ. Statisticians have long argued about the robustness of r as an estimate of ρ, but it would appear that r is only unbiassed if the sample data are normally distributed. In a review of this problem Kowalski states:

> "The general conclusion is that the distribution of r may be quite sensitive to non-normality, and the correlation analyses should be limited to situations in which (X,Y) is (at least very nearly) normal."
> (Kowalski; 1972, p.11)

Principal components analysis does not require the data to be normally distributed; the use of Pearson's r does. Norris (1971) discusses the effects of non-linear association between variables subjected to a components analysis, and concludes that erroneous interpretations of the structure of the data result from the use of the measure of linear association, r. Clark (1973) provides examples of the differences between component scores derived from "raw" (non-normal) data and those from transformed (approximately normal) data.

Further, if the data are normally distributed, there is a range of procedures available for hypothesis testing, including Bartlett's tests for the heterogeneity of the eigenvalues (section II(vii)), some of which are briefly mentioned in section IV(vi).

For the data used in Part III, the null hypotheses that the distribution of each of the variables was not different from the normal distribution with the same mean and variance were tested using the Kolmogorov-Smirnov one-sample test (Siegel; 1956, pp 47-52).* In no case, either before or after the exclusion of Dublin, could this hypothesis be rejected with even 95% confidence. However, if a χ^2 one-sample test (Siegel; 1956, pp 42-47) is used to test the null hypotheses that the distribution of each variable was not the same as a normal distribution with the same mean and variance, then for only one variable, "PASTURE", after the exclusion of Dublin, can this hypothesis be rejected with 95% confidence. In other words, the data are not significantly non-normal, but neither are they (with the exception of one variable) significantly normal. The exclusion of Dublin from the data set produced several χ^2 statistics that were much nearer to the rejection value than before, indicating that Dublin was a major cause of skew in the data set, and as was shown in section III(iii), the resultant distortion of the correlation matrix was so great as to produce the second component, accounting for 20% of the total variance.

---

* This, and all other hypothesis tests on the agricultural data, assumes that the 26 cases were independent and random samples drawn from all Irish counties! (But see the note by Cliff (1973; p.240)).

As Clark (1973, p.113) notes, most published analyses display a certain reticence about discussing the distribution of values for each variable. If it is not possible to transform skewed data to a normal distribution, then the use of either Spearman's rank correlation coefficient or Kendall's τ coefficient, both of which have up to 91% power efficiency (Siegel; 1956, p.223), is preferable to the use of Pearson's r. The use of rank correlation coefficients has largely been confined to the mental mappers, e.g. Gould and White (1974), whose data are usually obtained in ordinal form in the first place.

(ii) <u>Principal components analysis of dispersion matrices</u>

The principal components analysis of the variance-covariance matrix {C} is preferable to the analysis of the corresponding correlation matrix {R} for two reasons:

1. the sampling theory of the eigenvalues of dispersion matrices is more simple than that of correlation matrices, better understood, and thus there are more, and more exact, tests for the eigenvalues and eigenvectors of variance-covariance matrices;
2. the fact that the principal components are linear combinations of the original variables given by the general equation

$$y_{kj} = e_{1j}x_{k1} + e_{2j}x_{k2} + \ldots\ldots\ldots + e_{pj}x_{kp} \qquad (35)$$

where y_{kj} is the score of the k th data point on the j th component,

x_{ki} is the score of the k th data point on the i th variable,

and the e_{ij} are the elements of the matrix of normalised eigenvectors,

$(i = 1,2,\ldots,p\ ;\ j = 1,2,\ldots,p\ ;\ k = 1,2,\ldots,N)$

and that the x_{ki} still have the dimension in which they were measured, means that some physical interpretation of the e_{ij} may be possible, much as for the b coefficients in a regression analysis.

However, the measurement scales used in many geographical studies are so diverse that the disadvantages of not weighting each variable equally (by the conversion of the data to standard scores, and thus {C} to {R}) far exceed the advantages mentioned above. Although Walsh's data are all measured on the same scale, i.e. percentage change, the dispersion matrix was not analysed because of the large differences in variance between the variables, and also because, as percentage changes are dimensionless ratios, it was doubted if much physical interpretation would have been possible.

It would appear that the analysis of dispersion matrices is more suited to, and more commonly used on, problems in biology, psychology and economics.

(iii) <u>Discarding variables</u>

This section discusses two different, but related, problems. They are:

1. the choice of the variables determines the results of the principal components analysis - which variables should be chosen for analysis?
2. the principal components are one interpretation of the structure of intercorrelations between the variables - which variables best describe this structure, and thus could be most profitably monitored for further analysis?

An illustration of the first problem is the author's preliminary analysis of the soil moisture, topography and vegetation system of a savanna area in Brasil (Daultrey; 1970). Six of his twelve variables were direct measures of soil moisture, so it was hardly surprising that the first component, explaining 55% of the total variance, was interpreted on the basis of the component loadings as a "wet-dry" component. In a later analysis, three of the six soil moisture variables were discarded giving what the author considered to be a better description of the system, in that it was less biassed toward wet-dry contrasts (Daultrey; 1973). This was an entirely subjective decision, based on field knowledge of the system.

The selection of the 15 variables for analysis in Part III from the 39 derived by Walsh was again subjective, being based on an examination of the 39x39 correlation matrix, discarding those variables with very similar correlation patterns (i.e. those repeating the same information), and those with very weak correlations (i.e. those with very little information). Walsh's selections were different, and influenced by his far deeper knowledge of Irish agriculture than the author's. The results were similar - a first component relating to intensification and consolidation, and second and subsequent components produced by smaller, regional trends.

But generally speaking there is no solution to the first problem. If a set of highly intercorrelated variables is analysed, then a first component explaining a very large proportion of the total variance will be produced, which may be very pleasing but usually neither surprising nor edifying. If however, the variables are only weakly intercorrelated, the proportion of the variance explained by the first component will be very low, and it is quite likely that a large number of the other components will not be significantly different from each other in the proportion of the variance explained. (See also section IV(vi)).

A study of the second problem was recently carried out by Jolliffe (1972; 1973). He was concerned with methods of selection of a few indicator variables from a whole series so that these indicator variables adequately described the structure and variability of the system concerned. This has obvious practical advantages, for after a pilot study in which a large number of variables are examined, just the few indicator variables decided on could be monitored, giving a considerable saving in time and money spent on sampling and measurement.

He used three basic methods: multiple correlation: principal components analysis; and clustering of variables. In the first method, the variable with the highest multiple correlation coefficient, R, with all the others was discarded, and then the variable with the largest multiple correlation with the remaining variables, and so on until the largest R was less than a critical value, found to be 0.15. This would select those variables which were most nearly independent of each other.

Using principal components, two approaches were taken: for all components with λ_i less than a critical value λ_0, the variable which loaded highest on each component was discarded; or, for all components with λ_i greater than λ_0, the variable which loaded highest on each component was retained. The optimum value for λ_0 was found to be 0.7. The second approach, that of associating the variable most highly correlated with each of the largest components, seems more logical than the first, but the results produced are similar.

The third method involves clustering the variables into groups on the basis of their similarity to each other, and then selecting one variable from each group. These three methods all produced similar results, both with artificial and with real data.

One of the real data sets was that compiled by Moser and Scott (1961). They used a principal components analysis on 57 variables for 157 towns, and classified the towns according to the scores on the first four components (see section IV(v)). Jolliffe used the clustering method to reduce the 57 variables to 26, and produced a classification of towns similar to that of Moser and Scott. Jolliffe's analysis was obviously the cheaper in terms of computer time, but which of the two classifications was the better is still a matter for personal judgement.

Principal components analysis, then, can be used to study the correlation structure of multivariate observations by providing clues through the component loadings as to which of the variables best describe the independent trends in correlation. But no a priori assumptions as to the nature of the principal components can be made. Choosing variables to load onto components is a task for factor analysis.

(iv) <u>Rotation of principal components</u>

Factor analysts often rotate their factors so that loadings on certain variables are increased, and on others decreased, thus making more clear the interpretation of the factors. The principal components of a data set provide a unique orthogonal solution, maximising the variance explained by each successive component until the total variance has been accounted for. Often, factor analyses do not produce unique solutions, and never account for the total variance in a data set; hence the need for rotation. In principal components analysis there is little point in using an orthogonal rotation, because it eliminates the property of maximum variance; and little point in using a non-orthogonal rotation, because it eliminates the property of orthogonality. So there is neither the need for rotation, nor the justification. A debate on this issue, among several others, was conducted between Davies (1971a; 1971b; 1972a) and Mather (1971; 1972), and is well worth reading.

(v) <u>The analysis of component scores</u>

Having stressed the fact that principal components analysis is primarily a transformation procedure, it follows that the most useful application of the technique is not the interpretation of the components, but the further analysis of the component scores - the transformed data themselves. Before discussing this, it is necessary to examine one problem raised initially in section I(i). A few of the p components will account for most of the total variance. But how many components (and their scores) should be considered?

There is no easy answer, or one that is satisfactory to everyone. One obvious criterion is to eliminate all those components for which the null hypothesis $\lambda_1=\lambda_2=.....=\lambda_p$ cannot be rejected. When N is large, or when N is small but large relative to p, only the smallest components, if any, will not be heterogenous, leaving m acceptable components, m being not much less than p. A criterion used quite commonly is to eliminate all components whose eigenvalues are less than 1.0, on the grounds that these components are accounting for less of the total variance than any one variable (H.F. Kaiser, referenced in Pocock and Wishart; 1969, p.76). But when p is small, this may lead to the elimination of important dimensions of variability in the data set.

Most analyses report the cumulative proportion of the total variance accounted for by successive components. From this information, the number of components to be eliminated can be decided: either by retaining those that account for a previously determined proportion of the total variance, e.g. 80%, 90% (D.F. Morrison, referenced in Pocock and Wishart; 1969, p.77); or by discarding those components whose individual contribution to the proportion of total variance falls below a previously determined level, e.g. 10%, 5%. The advice given to the author as a student was to retain only those components which could be interpreted, on the grounds that to analyse scores on a component whose physical "meaning" was unclear was less than logical. However, as stated above, it is not the primary purpose of the technique to interpret the components, and anyway, analysis of the scores often gives a clue as to the "meaning" of the component.

In the author's experience, it doesn't really matter where the cut-off point is, provided the heterogeneity of the eigenvalues of the components can be demonstrated. The scores on the smaller components have smaller variances, thus their effect on the analysis is usually minimal. Again, the problem comes down to a subjective decision on the part of the analyst.

Because the scores are uncorrelated, they can be used as the m independent variables in a multiple regression on one dependent variable. This approach obviates the need to use stepwise multiple regression, with its often unsatisfactory criterion that the independent variable having the largest correlation with the dependent variable is the most "important" predictor of the dependent variable. But to interpret the regression coefficients, the principal components must be interpretable, otherwise the meaning of a linear combination of a linear combination of variables will be somewhat obscure. Lewis and Williams (1970), forecasting flood discharges, and Veitch (1970), forecasting maximum temperatures, have used principal component scores in this way, and Cox (1968) and Romsa et al. (1969) have likewise regressed scores derived from factor analyses.

Apart from simply mapping the scores, as in Part III, the most common type of analysis of scores is for them to be used in a hierarchical grouping procedure to produce a classification of the N cases based on m uncorrelated attributes rather than on p correlated attributes. The fact that the scores are uncorrelated allows very simple similarity coefficients (e.g. Euclidean distance) to be used in the grouping procedures (see, for example, Pocock and Wishart; 1969). The significance of the variation between the classes produced by the grouping procedure is usually tested by discriminant analysis or an analysis of variance (for a review, see Spence and Taylor; 1970).

In geographical studies the cases are more often than not areal units, thus the grouping of the cases produces areal aggregates, sometimes called regions. There is a vast literature on the quantitative approach to regionalisation, in which both principal components analysis and factor analytic techniques are used, and often confused. This general approach in human geography is summarised by Berry (1961; 1964; 1968) and more recently by Clark, Davies and Johnston (1974), and Mather and Openshaw (1974). Applied to urban areas the approach is often used as a part of "factorial ecology", or "social area analysis" and is discussed by Johnston (1971) and Rees (1971).

In physical geography the areal units of study are less often amenable to regionalisation by aggregation, although Mather and Doornkamp (1970) apply factor analysis, hierarchical grouping and discriminant analysis to classify a series of drainage basins. More often the classification of scores is used to locate boundaries in a continuum. For example McBoyle (1971) and Morgan (1971) have produced climatic classifications of Australia and West Malaysia respectively, and McCammon (1966) has distinguished depositional sub-environments in this way.

Clark, Davies and Johnston (1974, p.272) make the point that most analyses, in order to classify the cases, transform the variable space and then group the scores - this is an R-mode analysis. It may be more appropriate to transform the case space and group the cases by their loadings on the resultant components - i.e. a Q-mode analysis. Davies (1972b) provides an example of this approach in a study of journey patterns in the Swansea area. Conversely, botanists have traditionally used Q-mode principal components analysis to locate stands in species-dimensional space (e.g. Greig-Smith et al.; 1967), and more recently have used R-mode analysis to locate species in stand-dimensional space (e.g. Goff and Cottam; 1967). (Beals (1973) strongly criticises both uses of principal components analysis on the grounds that the botanical data do not meet the assumptions of the principal component model).

Q-mode analysis often produces the situation where there are more dimensions to be transformed than there are points in that space. Theoretically, this is not meaningful because, for example, a 3-dimensional plane cannot be defined uniquely by less than 3 points. Practically, this makes no difference, because the transformation is performed using individual estimates of the variances and covariances of the dimensions which are valid. However, no significance tests can be performed on the principal components so produced.

(vi) Sampling properties of principal components

Morrison (1967; section 7.7) provides a very comprehensible review of the more useful sampling properties of principal components, discussing confidence intervals for eigenvalues and eigenvectors derived from both the sample dispersion matrix and the sample correlation matrix, as well as tests for the null hypothesis that $\lambda_1=\lambda_2=.....=\lambda_p$. As mentioned in section II(vii), the Bartlett tests are not exact for the eigenvalues of correlation matrices. However, an exact test is available for the last p-1 roots of a correlation matrix, based on some general properties of the matrix.

It can be shown that the largest eigenvalue of a positive semidefinite covariance (or correlation) matrix cannot exceed the largest row sum of the absolute values (i.e. the sum of the values irrespective of sign).

$$\lambda_{max} \leq \sum_{j=1}^{p} |r_{ij}| \; max \tag{36}$$

This property is extremely useful, because an investigator can very quickly determine the largest proportion of the total variance that can be explained by any component.

If the correlation matrix $\{R\}$ is such that all the off-diagonal correlations are equal (an equicorrelation matrix), then the largest eigenvalue will equal the row sum

$$\lambda_{max} = 1 + (p - 1)r \tag{37}$$

and the remaining eigenvalues are all equal to $(1 - r)$.

Thus after the extraction of the first eigenvalue, the null hypothesis that the remaining eigenvalues are all equal can be tested by a χ^2 statistic relating to the heterogeneity of the original correlations.

The grand mean of the correlations, $\bar{r}$, is

$$\bar{r} = \frac{2}{p(p-1)} \cdot \left(\sum_{i>j} \sum r_{ij} \right) \tag{38}$$

The mean correlation of each variable with another, $\bar{r}_k$, is

$$\bar{r}_k = \frac{1}{p-1} \cdot \left(\sum_{\substack{j=1 \\ j \neq k}}^{p} r_{kj} \right) \tag{39}$$

The best estimate of the second and subsequent eigenvalues, $\hat{\lambda}$, if the matrix is an equicorrelation matrix, is

$$\hat{\lambda} = 1 - \bar{r} \tag{40}$$

Let $\hat{\mu}$ equal

$$\hat{\mu} = \frac{(p-1)^2(1 - \hat{\lambda}^2)}{p - (p-2)\hat{\lambda}^2} \tag{41}$$

χ^2 then becomes

$$\chi^2 = \frac{N-1}{\hat{\lambda}^2} \left(\left(\sum_{i>j} \sum (r_{ij} - \bar{r})^2 \right) - \hat{\mu} \left(\sum_{k=1}^{p} (\bar{r}_k - \bar{r})^2 \right) \right) \tag{42}$$

with $\frac{1}{2}(p + 1)(p - 2)$ degrees of freedom.

This test is exact if the population is multivariate normal, and N is large.

(vii) Conclusions

Because it is simple, easily accessible through package computer programs, and was introduced into geography early in the "quantitative revolution", principal components analysis has been much used, and often abused, by geographers. Only when used on carefully selected data (i.e. random, independent, and normally distributed) has it any "explanatory" power. The interpretation put on the results of the analyses in Part III was stretching the technique about as far as it can go. Principal components analysis is a transformation procedure, and should be used as such, or as a preliminary search procedure to identify variables (or cases) for further monitoring and analysis.

BIBLIOGRAPHY

A. Some Basic Textbooks

Anderson, T.W. (1958), *Introduction to multivariate statistical analysis*. (New York: John Wiley & Sons).

Goldstein, H. (1950), *Classical mechanics*. (Reading: Addison-Wesley Publ. Co).

Hammermesh, M. (1962), *Group theory and its application to physical problems*. (Oxford: Pergamon Press).

Harman, H. (1960), *Modern factor analysis*. (Chicago: University of Chicago Press).

Hope, K. (1968), *Methods of multivariate analysis*. (London: University of London Press).

Horst, P. (1965), *Factor analysis of data matrices*. (New York: Holt, Rinehart and Winston).

King, L.J. (1969), *Statistical analysis in geography*. (Englewood Cliffs, N.J.: Prentice-Hall).

Krumbein, W.C. & F.A. Graybill, (1965), *An introduction to statistical models in geology*. (New York: McGraw-Hill).

Lawley, D.N. & A.E. Maxwell, (1963), *Factor analysis as a statistical method*. (London: Butterworth).

Morrison, D.F. (1967), *Multivariate statistical methods*. (New York: McGraw-Hill).

Rummel, R.J. (1970), *Applied factor analysis*. (Evanston, Ill.: Northwestern University Press).

Siegel, S. (1956), *Nonparametric statistics for the behavioural sciences*. (New York: McGraw-Hill).

Thurstone, L.L. (1947), *Multiple-factor analysis*. (Chicago: University of Chicago Press).

B. Technical Considerations

Bartlett, M.S. (1950), Tests of significance in factor analysis. *British Journal of Statistical Psychology*, 3, 77-85.

Bartlett, M.S. (1951a), The effect of standardisation on a approximation in factor analysis. *Biometrika*, 38, 337-344.

Bartlett, M.S. (1951b), A further note on tests of significance in factor analysis. *British Journal of Statistical Psychology*, 4, 1-2.

Beals, E.W. (1973), Ordination: mathematical elegance and ecological naivete. *Journal of Ecology*, 61, 23-35.

Clark, D. (1973), Normality, transformation, and the principal components solution: an empirical note. *Area*, 5, 110-113.

Cliff, A.D. (1973), A note on statistical hypothesis testing. *Area*, 5, 240.

Daultrey, S. (1973), Some problems in the use of principal components analysis. IBG Quantitative Methods Study Group, *Working Papers*, 1, 13-18.

Davies, W.K.D. (1971a), Varimax and the destruction of generality: a methodological note. *Area*, 3, 112-118.

Davies, W.K.D. (1971b), Varimax and generality: a reply. *Area*, 3, 254-259.

Davies, W.K.D. (1972a), Varimax and generality: a second reply. *Area*, 4, 207-208.

Goff, F.G. & G. Cottam, (1967), Gradient analysis: the use of species and synthetic indices. *Ecology*, 48, 783-806.

Hotelling, H. (1933), Analysis of a complex of statistical variables into principal components. *Journal of Educational Psychology*, 24, 417-441, 498-520.

Johnston, R.J. (1971), Some limitations on factorial ecologies and social area analysis. *Economic Geography*, 47, 314-323.

Jolliffe, I.J. (1972), Discarding variables in a principal component analysis. I: Artificial data. *Applied Statistics*, 21, 160-173.

Jolliffe, I.J. (1973), Discarding variables in a principal component analysis. II: Real data. *Applied Statistics*, 22, 21-31.

Kowalski, C.J. (1972), On the effects of non-normality on the distribution of the sample product-moment correlation-coefficient. *Applied Statistics*, 21, 1-12.

Mather, P.M. (1971), Varimax and generality: a comment. *Area*, 3, 252-254.

Mather, P.M. (1972), Varimax and generality: a second comment. *Area*, 4, 27-30.

Mather, P.M. & S. Openshaw, (1974), Multivariate methods and geographical data. *The Statistician*, 23, 283-308.

Norris, J.M. (1971), Functional relationships in the interpretation of principal components analyses. *Area*, 3, 217-220.

C. Applications

Berry, B.J.L. (1961), 'A method for deriving multi-factor uniform regions'. *Przeglad Geograficzne*, 33, 263-279.

Berry, B.J.L. (1964), Approaches to regional analysis: a synthesis. *Annals, Association of American Geographers*, 54, 2-11.

Berry, B.J.L. (1968), A synthesis of formal and functional regions using a general field theory of spatial behaviour'. in B.J.L. Berry and D.F. Marble, eds, *Spatial analysis: a reader in statistical geography*. (Englewood Cliffs, N.J.: Prentice-Hall).

Clark, D., W.K.D. Davies & R.J. Johnston, (1974), 'The application of factor analysis in human geography'. *The Statistician*, 23, 259-281.

Cox, K.R. (1968), 'Suburbia and voting behaviour in the London metropolitan area. *Annals, Association of American Geographers*, 58, 111-127.

Daultrey, S. (1970), An analysis of the relation between soil moisture, topography, and vegetation types in a savanna area'. *Geographical Journal*, 136, 399-406.

Davies, W.K.D. (1972b), Conurbation and city region in an administrative borderland: a case study of the greater Swansea area. *Regional Studies,* 6, 217-236.

Gould, P.R. (1967), On the geographical interpretation of eigenvalues. *Transactions Institute of British Geographers,* 42, 53-86.

Gould, P.R. & R. White, (1974), *Mental maps,* (Harmondsworth: Penguin).

Greig-Smith, P., M.P. Austin & T.C. Whitmore, (1967), The applications of quantitative methods to vegetation survey. I: Association analysis and principal component ordinations of rain forest. *Journal of Ecology,* 55, 483-503.

Lewis, G.L. & T.T. Williams, (1970), Selected multivariate statistical methods applied to runoff data from Montana watersheds. *State Highway Commission and U.S. Bureau of Roads, Publ. P3 183705.*

McBoyle, G. (1971), Climatic classification of Australia by computer. *Australian Geographical Studies,* 9, 1-14.

McCamman, R.B. (1966), Principal components analysis and its application in large-scale correlation studies. *Journal of Geology,* 74, 721-733.

Mather, P.M. & J.C. Doornkamp, (1970), Multivariate analysis in geography with particular reference to drainage-basin morphology. *Transactions Institute of British Geographers,* 51, 163-187.

Morgan, R.P.C. (1971), Rainfall of West Malaysia - a preliminary regionalisation using principal components analysis. *Area,* 3, 222-227.

Moser, C.A. & W. Scott, (1961), *British towns.* Edinburgh: Oliver and Boyd.

Pocock, D.C.D. & D. Wishart, (1969), 'Methods of deriving multi-factor uniform regions. *Transactions Institute of British Geographers,* 47, 73-98.

Rees, P.H. (1971), Factorial ecology: an extended definition, survey, and critique. *Economic Geography,* 47, 220-233.

Romsa, G.H., W.L. Hoffman, S.T. Gladin & S.D. Brunn, (1969), An example of the factor analytic - regression model in geographic research. *Professional Geographer,* 21, 344-346.

Spence, N.A. & P.J. Taylor, (1970), Quantitative methods in regional taxonomy. in C. Board et. al (eds) *Progress in Geography, Volume 2,* 1-63 (London: Arnold).

Veitch, L.G. (1970), Forecasting Adelaide's maximum temperature using principal components and multiple linear regression. *Australian Meteorological Magazine,* 18, 1-21.

Walsh, J. (1974), *Changes in Irish agriculture.* unpublished B.A. dissertation, Dept. of Geography, University College Dublin.

NOTES

JUL